SuperSymmetry and the Unified SuperStandard Model

Stephen Blaha Ph. D.
Blaha Research

Pingree-Hill Publishing

To Margaret

Some Other Books by Stephen Blaha

All the Megaverse! Starships Exploring the Endless Universes of the Cosmos using the Baryonic Force (Blaha Research, Auburn, NH, 2014)

SuperCivilizations: Civilizations as Superorganisms (McMann-Fisher Publishing, Auburn, NH, 2010)

PHYSICS IS LOGIC PAINTED ON THE VOID: Origin of Bare Masses and The Standard Model in Logic, U(4) Origin of the Generations, Normal and Dark Baryonic Forces, Dark Matter, Dark Energy, The Big Bang, Complex General Relativity, A Megaverse of Universe Particles (Blaha Research, Auburn, NH, 2015).

The Origin of Higgs ("God") Particles and the Higgs Mechanism: Physics is Logic III, Beyond Higgs – A Revamped Theory With a Local Arrow of Time, The Theory of Everything Enhanced, Why Inertial Frames are Special, Universes of the Mind (Blaha Research, Auburn, NH, 2015).

New Types of Dark Matter, Big Bang Equipartition, and A New U(4) Symmetry in the Theory of Everything: Equipartition Principle for Fermions, Matter is 83.33% Dark, Penetrating the Veil of the Big Bang, Explicit QFT Quark Confinement and Charmonium, Physics is Logic V (Blaha Research, Auburn, NH, 2015).

The Periodic Table of the 192 Quarks and Leptons in The Theory of Everything: The U(4) Layer Group, Physics is Logic VI (Blaha Research, Auburn, NH, 2015).

New Boson Quantum Field Theory, Dark Matter Dynamics, Dark Matter Fermion Layer Mixing, Genesis of Higgs Particles, New Layer Higgs Masses, Higgs Coupling Constants, Non-Abelian Higgs Gauge Fields, Physics is Logic VII (Blaha Research, Auburn, NH, 2015)

CQMechanics: A Unification of Quantum & Classical Mechanics, Quantum/Semi-Classical Entanglement, Quantum/Classical Path Integrals, Quantum/Classical Chaos (Blaha Research, Auburn, NH, 2016).

All the Universe! Faster Than Light Tachyon Quark Starships & Particle Accelerators with the LHC as a Prototype Starship Drive Scientific Edition (Pingree-Hill Publishing, Auburn, NH, 2011).

From Asynchronous Logic to The Standard Model to Superflight to the Stars; Volume 2: Superluminal CP and CPT, U(4) Complex General Relativity and The Standard Model, Complex Vierbein General Relativity, Kinetic Theory, Thermodynamics (Blaha Research, Auburn, NH, 2012)

New Boson Quantum Field Theory, Dark Matter Dynamics, Dark Matter Fermion Layer Mixing, Genesis of Higgs Particles, New Layer Higgs Masses, Higgs Coupling Constants, Non-Abelian Higgs Gauge Fields, Physics is Logic VII (Blaha Research, Auburn, NH, 2015)

The Origin of Fermions and Bosons, and Their Unification (Pingree-Hill Publishing, Auburn, NH, 2017).

Megaverse: The Universe of Universes (Pingree Hill Publishing, Auburn, NH, 2017).

Available on bn.com, Amazon.com, Amazon.co.uk and other international web sites as well as at better bookstores (through Ingram Distributors).

CONTENTS

INTRODUCTION

The author has been engaged in a seventeen year effort to build a unified theory of elementary particles and gravitation based on Quantum Field Theory, suitably extended to make it finite, and having asymptotic states that are well-defined in any coordinate system. Quantum Field Theory is without a doubt the most accurate theoretic framework that we have found. There is direct evidence for this claim in its ability to account for electromagnetic phenomena to incredible accuracy using Quantum Electrodynamics. No other scientific theory has achieved the degree of precision seen in the agreement of experiment and theory in fundamental electromagnetic phenomena.

This book describes a further extension of the author's unified Extended Standard Model to a Unified SuperStandard Model of Elementary Particles and Gravitation. It develops a new formulation of the Layer group that creates four layers of four generations of fermions. It also creates four layers of vector bosons. The number of fermions and of vector bosons is 192.

Based on the Dark and normal fermions and bosons of the Unified SuperStandard Model, and an Equipartition Principle, we find the proportion of Dark Matter was 83%; and the proportion of Dark mass-energy was 91% at the Big Bang point. These estimates are close to today's measured values.

The book shows how a form of SuperSymmetry results, in which there is a U(192) group that 'rotates' vector bosons and fermions. We show it is possible to also define fermionic operators (numbering 192^2 operators) that can transform between vector bosons and fermions. When the theory is extended from 4-dimensional space-time to 192-dimensional space (the *Megaverse*), the U(192) group can then be extended to apply to Megaverse coordinates as well as fermions and vector bosons. Thus the U(192) symmetry group, which we call Θ-symmetry, becomes a 'full-fledged' symmetry group in Megaverse space. In addition to this group, we also define a Complex Lorentz group with one time coordinate and 191 spatial coordinates.

Higher dimensional complex spaces, the lack of finiteness in perturbation theory calculations, and the need to have a form of quantum field theory that preserves particle numbers in asymptotic states, all lead to a 'new' more general form of quantum theory based on PseudoQuantum Field Theory, and Two-Tier quantum coordinates.

We show, in detail, that elementary particle theory is analogous to quantum finite state machines. We reinforce the concept that elementary particle physics and gravitation can be logically derived from fundamental principles in the manner of Euclidean geometry. We conclude by showing that physics only makes sense if there is an 'Unmoved Mover' as proposed by ancient and modern philosophers.

1. Blaha's Unified Extended Standard Model

In this chapter we outline our unified Extended Standard Model of elementary particles and gravitation which was derived directly from Special and General Relativity.[1] Our theory is described in detail in *The Origin of Fermions and Bosons, And Their Unification* (referred to as volume I in this book), which appeared earlier in 2017 as well as in earlier books. We are describimg the theory presented in volume I and earlier books, for the purpose of comparing it to our new, more general, theory that we will present in the next chapter. The new theory appears to have a remarkable set of new features.

1.1 Fundamental Unified Extended Model Theory

Volume I takes the Complex Lorentz group in flat space-time as its starting point and derives the four types (species) of fermions directly: charged lepton, uncharged lepton, up-type quark and down-type quark. It then proceeds to show that the interactions of the Standard Model $SU(3)\otimes SU(2)\otimes U(1)$ follow directly from Complex Lorentz group geometry. We call this group the Reality group[2] beause it maps complex coordinate systems, generated by complex Lorentz transformations, to real-valued coordinate systems. The existence of Dark Matter leads to an enlargement of the Reality group to $SU(3)\otimes SU(2)\otimes U(1)\otimes SU(2)\otimes U(1)$ with the additional $SU(2)\otimes U(1)$ factors providing the interactions of Dark Matter. We do not include a Dark Matter strong interaction because 1) if Dark Matter had the known SU(3) Strong Interaction then it wouldn't be 'Dark', and 2) the group $R = SU(3)\otimes SU(2)\otimes U(1)\otimes SU(2)\otimes U(1)$ has 16 generators that are all that is needed to transform any flat space-time complex-valued coordinate system to a real-valued coordinate system. Thus Complex Lorentz group space-time geometry requires R and only R—additional Reality factors are superfluous.

Having established the group structure, fermion spectrum, and specified interactions for the one generation case, we then proceeded to introduce fermion generations. We began by noting that there exist particle number operators: Baryon number, Lepton number, and their Dark equivalents that are conserved according to known experimental data. These four number operators lead directly to a U(4) Generation group that generates four generations of each species of fermion (although only three generations are known at present).

The existence of four generations of each species then leads in turn to another U(4) group—the Layer group[3]—that gives us a Periodic Table of Fermions with three additional

[1] Our theory is described in detail in Blaha (2017b) which is denoted I in this book.

[2] This group does not combine space-time and internal symmetries avoiding the Coleman-Mandula Theorem.

[3] Discussed later in detail for our Extended Standard Model. In chapter 2 we define a new form of Layer group.

replicas of the 'known' layer of four generations (with changes in other layer fermion masses.) In total we find there are 192 distinct fundamental fermions – 128 quarks and 64 leptons.

Having discovered the riches of the Complex Lorentz group as it leads us from four fermion species to 192 different fermions, we then realize that consistency compels us to accept Complex General Relativity as well (with the Complex Lorentz group as its flat space-time limit.) It would be illogical if General Relativity was real-valued and flat space-time were complex-valued. So volume I proceeded to consider Complex General Relativity and discovered that each Complex General Coordinate Transformations can be factored into a real General Coordinate transformation and a complex coordinate transformation. This factorization leads to another U(4) group—that we called the General Relativistic Reality group. This group is distinct from the Complex Lorentz group Reality group R. This new group adds a new set of interactions that affects fermion and boson masses through the Higgs Mechanism. By adding a mass term to each fundamental particle's mass it sets the universe mass scale and yields the principle of equality of inertial and gravitational mass without appeal to Mach or others.

Taken altogether volume I creates a unified theory of an Extended Standard Model and General Relativity with the symmetry $SU(3) \otimes SU(2) \otimes U(1) \otimes SU(2) \otimes U(1) \otimes U(4) \otimes U(4) \otimes U(4)$ plus real-valued General Coordinate transformations. Given this menagerie of interactions volume I develops a formalism for the 'rotation of interactions' called Ω-symmetry, which introduces a new group symmetry, $SU(3) \otimes U(64)$, and a gauge field providing interactions between all 192 fermions. This (broken) symmetry group's transformations rotate the eight particle symmetries[4] amongst each other, and, correspondingly rotate the 192 fermions amongst each other. The Weinberg rotation of ElectroWeak theory is a simple example of this form of interaction rotation. Thus the theory is invariant under this symmetry although it is broken to $SU(3) \otimes SU(3) \otimes U(56)$ as shown in volume I.[5]

This 'rotation of rotations' symmetry group completes the unification process in our unified Extended Standard Model describe in volume I. For if one can rotate things into each other, they are, in a sense, manifestations of the same unified structure.

The unified theory presented in volume I has a naturalness, and simplicity, that strongly indicate it is on the right track. In chapter 2 we modify the unified theory to obtain a more 'perfect' unified physical theory.

We can summarize the theory of volume I as: From the Complex Lorentz group it extracts the four species of fermions; it grows the four species to four generations based on Baryon and Lepton number conservation (perhaps slightly broken) which yield the U(4) Generation group; the four generations lead to four (broken) conservation laws that in turn yield four layers of four generations of fermions. Complex General Relativity completes the initial development with another U(4) interactions group. Particle masses are generated by the Higgs

[4] These symmetry groups are $SU(3)_{Color}$, $SU(2)_{Weak}$, $U(1)_{Weak}$, , $SU(2)_{DarkWeak}$, $U(1)_{DarkWeak}$, $U(4)_{Generation}$, $U(4)_{Layer}$, and $U(4)_{RealGrav}$ in the notation of I.

[5] In chapter 2 we describe a new form of the 'rotation of interactions' group.

Mechanism. Then, unifying the interactions, we obtain a unified theory with interactions rotations.

Volume I concludes by showing that the gravity potential found experimentally at solar system distances, at galactic distances, and at inter-galactic distances follow from our theory. It also shows that the theory yields a linear quark potential and the Charmonium potential. Lastly it shows the theory may explain the missing proton spin puzzle, and also the difference in the proton radius of the hydrogen and muonic hydrogen atoms.

1.2 Fermion Periodic Table

The Periodic Table of 192 Fermions in volume I was constructed based on a number of reasonable (in the author's view) assumptions.[6] First we required that the fermion species (types) be constructed by complex Lorentz group transformations on a Dirac spinor at rest. This yielded four species of fermions that we identified with charged leptons, neutral leptons, up-type quarks, and down-type quarks. We then found that quarks occupied the $\underline{3}$ of color SU(3). Color SU(3) as well as the normal and Dark Weak interactions of group SU(2), and the normal and Dark electromagnetic interactions were 'derived' from the form of a Lorentz transformation.[7] Thus we found eight species of normal fermions, and four species of Dark fermions (since Dark fermions do not have color interactions – they are SU(3) singlets.) Thus in one generation there are eight normal fermions and four Dark fermions.

Then noting that there are four conserved particle number operators (baryon number,[8] lepton number, Dark baryon number and Dark lepton number) we were led to posit a U(4) Generation symmetry that generated four generations of fermions. Thus we found 32 normal fermions and 16 Dark fermions = 48 fermions in the four generations.

Next we noted that the number of particles in each of the four generations was (almost) conserved. This fact led to another group – the U(4) Layer group – that led to four layers of fermions. The total number of fermions then was found to be 4*48 = 192 fermions (Fig. 1.1).

1.3 Boson Interactions

The eleven interactions[9] that we identified in volume I, and in earlier books, are

[6] The contents of this, and the following, sections, and the figure, appear in several of the author's earlier books in 2015 and 2016.

[7] These points all appear in Blaha (2015a).

[8] The author has pointed out in previous books that disparities in measurements of the gravitational constant G could reflect the existence of a baryonic force that, like electromagnetism, leads to a conserved baryon number. Similar comments would apply to other conserved Generation group numbers and almost conserved Layer group numbers.

[9] In voume I, and in earlier books, we use a Pseudoquantum quantum field theory formalism that associates a second field with each interaction. In the interests of simplicity we use the usual one field formalism in this section.

1.3.1 The SU(2) Weak Interaction

The Weak interaction SU(2) gauge field is defined as $W^{i\mu}(x)$ for i = 1, 2, 3. Using SU(2) generators we define the matrix form by $W^{\mu}(x) = W^{i\mu}(x)\tau_i$.

1.3.2 The U(1) Electromagnetic Interaction

The U(1) electromagnetic gauge field[10] is defined as $A_E^{\mu}(x)$.

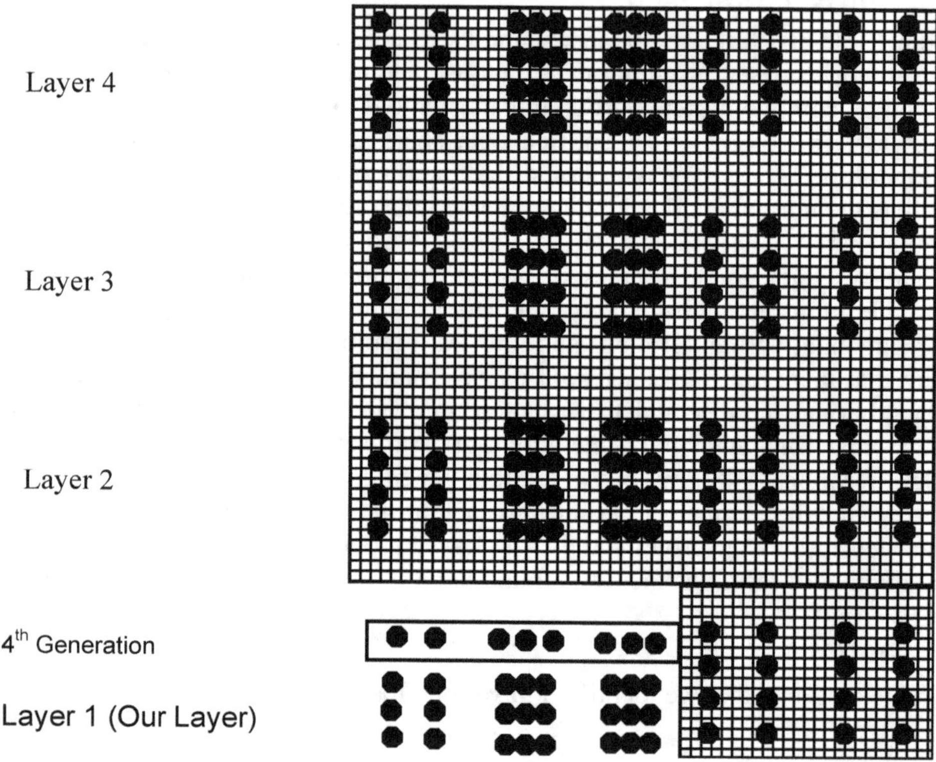

Figure 1.1. Dark parts of the periodic table are 'cross-hatched.' Light parts are the known fermions – with an additional, as yet not found, 4th generation of layer 1 is shown boxed. It is part of 'Dark matter' at present. When found experimentally it will be 'non-Dark.' Normal quarks appear as triplets. Dark quarks are singlets.

[10] We separate the ElectroWeak theory into a Weak SU(2) part and an electromagnetic U(1) part. They can be united through an inverse ElectroWeak rotation.

1.3.3 The SU(3) Strong Interaction

The Strong SU(3) gauge field is defined as $A_{SU(3)}{}^{i\mu}(x)$ for i = 1, ... , 8. Using SU(3) generators we define the matrix form by $A_{SU(3)}{}^{\mu}(x) = A_{SU(3)}{}^{i\mu}(x)T_i$.

1.3.4 The U(4) Generation Group Interaction

The U(4) Generation group[11] matrix generators are denoted G_i and its gauge fields are denoted $U^i_\mu(X)$. Thus the Generation group terms in covariant derivatives are

$$\mathbf{U^i_\mu \cdot G_i}$$

where i = 1, ... , 16. The Generation group is described in detail in chapters 12 and 13 of Blaha (2017b), and in earlier books by the author.

1.3.5 The U(4) Layer Group Interaction

The U(4) Layer group[12] matrix generators are denoted G_{Lk} and its gauge fields are denoted $V^i_\mu(X)$. Thus the Layer group terms in covariant derivatives are

$$\mathbf{V^i_\mu \cdot G_{Li}}$$

where i = 1, 2, ..., 16. The Layer group is described in chapters 15, 16 and 17 of Blaha (2017b), and in earlier books by the author.

1.3.6 The SU(2) Dark Weak Interaction

We assume Dark Weak interactions have the same form as the known SU(2) Weak interactions. the The Dark Weak interaction SU(2) gauge field is defined as $W_D{}^{i\mu}(x)$ for i = 1, 2, 3. Using SU(2) matrix generators we define the matrix form by $W_D{}^{\mu}(x) = W_D{}^{i\mu}(x)\tau_{Di}$ where the generator matrices τ_{Di} are not in the same subspace as the normal SU(2) generators.

1.3.7 The U(1) Dark Electromagnetic Interaction

The U(1) Dark electromagnetic gauge field[13] is defined as $A_{DE}{}^{\mu}(x)$.

1.3.8 The U(4) General Relativistic Reality Group Interaction – The Species Group

The U(4) Species group (the General Relativistic Reality Group) interaction gauge field[14] is $A_S{}^{\mu}(x) = A_{R_{flat}}{}^{\mu}(x)$. This group rotates fermion fields amongst the four normal species

[11] If there are only three generatons of fermions then the Generation group is U(3).

[12] If there are only three generatons of fermions then the Layer group is also U(3).

[13] We introduce two fields as we did in our article S. Blaha, Phys. Rev. D**10**, 4268 (July, 1974). These fields enable us to define a free electromagnetic lagrangian that is linear in the fields for reasons given elsewhere.

and the four Dark species. Each Color subspecies is rotated to a color species of the same color. Intermediate rotations fall into one species or another. See chapter 23 of volume I for more details.

1.3.9 The 'Interaction Rotation' Interaction - A_Ω

The SU3)⊗U(64) interaction rotation group gauge field[15] is defined as $A_\Omega{}^{ij\mu}(x)$ for $i = 1, ..., 8$ and $j = 1, ..., 64$. They total 512 gauge field components. Using their 72 generators expressed in the SU(3) $\underline{3}$ representation and the U(64) $\underline{64}$ representation, with matrix denoted $T_{\Omega ij}$, we can define the matrix form as

$$A_\Omega{}^\mu(x) = A_\Omega{}^{ij\mu}(x)T_{\Omega ij}$$

where $T_{\Omega ij}$ is a cross product of SU3)⊗U(64) generators. The tensor product generator matrices are 192×192 matrices (since 3*64 = 192). We choose to have a representation with 192×192 matrices due to the 192 fermions in our Periodic Table of Fermions. (See chapter 16 of volume I.)

This interaction, which we will call the *Ω-interaction*, is described in detail in chapter 31 of volume I. The gauge fields will correspondingly be called *Ω-fields*.

1.3.10 The Spinor Connection Interaction

The spinor connection used in formulations of vierbein gravity is $B_{\mu ab}(x)$ where a and b are tangent space indices. The vector is combined with γ matrices for use in matrix equations:

$$B^\mu = B^\mu{}_{ab}\Sigma^{ab}$$

where

$$\Sigma^{ab} = i\,[\gamma^a, \gamma^b]/4$$

Under a local Lorentz transformation S

$$B^\mu(x) \to S(x)B^\mu(x)S^{-1}(x) - i\,S(x)\partial^\mu S^{-1}(x)$$

A simple spin ½ field transforms as

$$(\partial^\mu + i\,B^{1\mu} + i\,B^{2\mu})\psi \to S(\partial^\mu + i\,B^{1\mu} + i\,B^{2\mu})\psi$$

1.3.11 Real-Valued General Coordinate Connection (Interaction)

The usual gravitational metric field $g_{\mu\nu}$ with

[14] See chapters 22 and 23 of volume I for a discussion of the origin of this interaction in Complex General Relativity.
[15] The SU(3) factor is *not* color SU(3).

$$\Gamma_{GR}{}^{\lambda}{}_{\mu\nu} = \tfrac{1}{2}g^{\lambda\alpha}(\partial_\mu g_{\alpha\nu} + \partial_\nu g_{\alpha\mu} - \partial_\alpha g_{\mu\nu})$$

1.4 The Covariant Derivative

The covariant derivative[16] which appears in fermion and gravitation equations uses

$$\mathbf{A}_I{}^{\mu} = (g_1\mathbf{A}_{SU(3)}{}^{\mu}(x_C),\ g_2\mathbf{W}^{\mu}(x)\ ,\ g_3\mathbf{A}_E{}^{\mu}(x),\ g_4\mathbf{W}_D{}^{\mu}(x),\ g_5\mathbf{A}_{DE}{}^{\mu}(x),\ g_6\mathbf{U}^{\mu}(x),\ g_7\mathbf{V}^{\mu}(x),\ g_8\mathbf{A}_S{}^{\mu}(x))$$

$$(19.16)$$

where each element is a vector of the gauge fields of the respective group. For simplicity we label the respective coupling constants as g_1, g_2, ... , g_8. In eq. 19.16 above the subscript 'D' labels Dark matter interactions, 'W' labels Weak fields, 'E' labels Electromagnetic fields, $V^{\mu}(x)$ labels U(4) Generation group fields, and 'V' labels U(4) Layer group fields. A_S labels the U(4) General Relativistic transformations Reality field.

The interactions' symmetry is $SU(3)\otimes SU(2)\otimes U(1)\otimes SU(2)\otimes U(1)\otimes U(4)\otimes U(4)\otimes U(4)$ respectively. The number of gauge fields for each of the elements of the vectors is 8. 3, 1, 3, 1, 16, 16, and 16 respectively – totaling 64 fields.

Similarly we define an 8-vector of 64 matrix generators

$$\mathbf{T}_I = (\mathbf{T}_{SU(3)},\ \tau_{SU(2)},\ \mathbf{I}_{U(1)},\ \tau_{DSU(2)},\ \mathbf{I}_{DU(1)},\ \mathbf{G}_{U(4)},\ \mathbf{G}_{LU(4)},\ \mathbf{G}_S) \qquad (19.17)$$

Note that the $SU(3)\otimes SU(2)\otimes U(1)\otimes SU(2)\otimes U(1)\otimes U(4)\otimes U(4)$ fields are the same in all four layers in the Extended Standard Model described in volume I. Since higher layer fermions have not as yet been found experimentally we see that there are two possible resolutions of this issue:

1. Each layer has a different set of these fields, so that direct interactions between layers do not exist through these fields. We discuss this possibility in more detail in chapter 2.

2. Higher layer fermions are so massive that they are not accessible in current accelerators.

Using the above definitions the covariant derivative of a 4-vector is

$$\begin{aligned}
D_\nu V_\mu &= (\partial_\nu + iF_\nu)V_\mu - H^{\sigma}{}_{\nu\mu}V_\sigma \\
&= [g^{\sigma}{}_\mu\partial_\nu + ig^{\sigma}{}_\mu F_\nu - H^{\sigma}{}_{\nu\mu}]V_\sigma \\
&= [g^{\sigma}{}_\mu\partial_\nu + iD^{\sigma}{}_{\mu\nu}]V_\sigma
\end{aligned} \qquad (26.2)$$

where[17]

$$F^{\mu} = g_\Omega\mathbf{A}_\Omega{}^{1\mu}(x) + g_\Omega\mathbf{A}_\Omega{}^{2\mu}(x) + \mathbf{A}_I{}^{1\mu}(x) + \mathbf{A}_I{}^{2\mu}(x) + B^{1\mu} + B^{2\mu} \qquad (26.3)$$

[16] This section has equations obtained from volume I. The equation numbering is that of volume I.
[17] We will omit the insertion of coupling constants of $B^{1\mu}$ and $B^{2\mu}$ in the interests of simplifying expressions.

by eqs. 19.16 and 19.17, and

$$H^\sigma{}_{\nu\mu} = \Gamma_{GR}{}^\sigma{}_{\nu\mu} + \Gamma_{GR}{}^{2\sigma}{}_{\nu\mu} \tag{26.4}$$

$$D^\sigma{}_{\mu\nu} = g^\sigma{}_\mu F_\nu + iH^\sigma{}_{\nu\mu} \tag{26.5}$$

where we have abstracted the complex part of the complex affine connection into the U(4) gauge field $A_S{}^\mu$. Eq. 26.4 is the real-valued part of the complex affine connection.

Commutators of the vector fields in F_μ are implicit when the covariant derivative is applied to vectors and tensors such as V_σ.

1.5 The Curvature Tensor

The curvature tensor (applied to a 4-vector) defined in volume I is

$$R'^\beta{}_{\sigma\nu\mu}V_\beta = g^\alpha{}_\mu(\partial_\nu + iF_\nu)g^\beta{}_\sigma(\partial_\alpha + iF_\alpha)V_\beta - H^\alpha{}_{\mu\nu}g^\beta{}_\sigma(\partial_\alpha + iF_\alpha)V_\beta +$$
$$+ H^\alpha{}_{\mu\nu}H^\beta{}_{\sigma\alpha}V_\beta - g^\alpha{}_\mu(\partial_\nu + iF_\nu)H^\beta{}_{\sigma\alpha}V_\beta - H^\gamma{}_{\nu\sigma}\{g^\alpha{}_\gamma(\partial_\mu + iF_\mu)V_\alpha - H^\alpha{}_{\gamma\mu}V_\alpha\} -$$
$$- \{\mu \leftrightarrow \nu\}$$

$$= ig^\beta{}_\sigma(\partial_\nu F_\mu - \partial_\mu F_\nu - i[F_\nu, F_\mu])V_\beta + (\partial_\mu H^\beta{}_{\sigma\nu} - \partial_\nu H^\beta{}_{\sigma\mu} + H^\gamma{}_{\nu\sigma}H^\beta{}_{\gamma\mu} - H^\gamma{}_{\mu\sigma}H^\beta{}_{\gamma\nu})V_\beta$$

$$= ig^\beta{}_\sigma(F_E{}^1{}_{\nu\mu} + F_E{}^2{}_{\nu\mu} + F_W{}^1{}_{\nu\mu} + F_W{}^2{}_{\nu\mu} + F_{DE}{}^1{}_{\nu\mu} + F_{DE}{}^2{}_{\nu\mu} + F_{DW}{}^1{}_{\nu\mu} + F_{DW}{}^2{}_{\nu\mu} + F_{SU(3)}{}^1{}_{\nu\mu} +$$

$$+ F_{SU(3)}{}^2{}_{\nu\mu} + F_U{}^1{}_{\nu\mu} + F_U{}^2{}_{\nu\mu} + F_V{}^1{}_{\nu\mu} + F_V{}^2{}_{\nu\mu} + F_\Omega{}^1{}_{\nu\mu} + F_\Omega{}^2{}_{\nu\mu})V_\beta +$$
$$+ (ig^\beta{}_\sigma B^1{}_{\nu\mu} + ig^\beta{}_\sigma B^2{}_{\nu\mu} + \partial_\mu H^\beta{}_{\sigma\nu} - \partial_\nu H^\beta{}_{\sigma\mu} + H^\gamma{}_{\nu\sigma}H^\beta{}_{\gamma\mu} - H^\gamma{}_{\mu\sigma}H^\beta{}_{\gamma\nu})V_\beta$$

$$= R'_E{}^\beta{}_{\sigma\nu\mu}V_\beta + R'_{SU(2)}{}^\beta{}_{\sigma\nu\mu}V_\beta + R'_{DE}{}^\beta{}_{\sigma\nu\mu}V_\beta + R'_{DSU(2)}{}^\beta{}_{\sigma\nu\mu}V_\beta + R'_{SU(3)}{}^\beta{}_{\sigma\nu\mu}V_\beta + R'_U{}^\beta{}_{\sigma\nu\mu}V_\beta +$$
$$+ R'_V{}^\beta{}_{\sigma\nu\mu}V_\beta + R'_S{}^\beta{}_{\sigma\nu}V_\beta + R'_\Omega{}^\beta{}_{\sigma\nu}V_\beta + R'_B{}^\beta{}_{\sigma\nu}V_\beta + R'_G{}^\beta{}_{\sigma\nu\mu}V_\beta \tag{26.8}$$

where

$$R'_{SU(3)}{}^\beta{}_{\sigma\nu\mu} = ig^\beta{}_\sigma(F_{SU(3)}{}^1{}_{\nu\mu} + F_{SU(3)}{}^2{}_{\nu\mu}) \tag{26.9}$$
$$R'_{SU(2)}{}^\beta{}_{\sigma\nu\mu} = ig^\beta{}_\sigma(F_W{}^1{}_{\nu\mu} + F_W{}^2{}_{\nu\mu})$$

$$R'_E{}^\beta{}_{\sigma\nu\mu} = ig^\beta{}_\sigma(F_E{}^1{}_{\nu\mu} + F_E{}^2{}_{\nu\mu})$$
$$R'_U{}^\beta{}_{\sigma\nu\mu} = ig^\beta{}_\sigma(F_U{}^1{}_{\nu\mu} + F_U{}^2{}_{\nu\mu})$$
$$R'_V{}^\beta{}_{\sigma\nu\mu} = ig^\beta{}_\sigma(F_V{}^1{}_{\nu\mu} + F_V{}^2{}_{\nu\mu})$$
$$R'_{DSU(2)}{}^\beta{}_{\sigma\nu\mu} = ig^\beta{}_\sigma(F_{DW}{}^1{}_{\nu\mu} + F_{DW}{}^2{}_{\nu\mu})$$
$$R'_{DE}{}^\beta{}_{\sigma\nu\mu} = ig^\beta{}_\sigma(F_{DE}{}^1{}_{\nu\mu} + F_{DE}{}^2{}_{\nu\mu})$$
$$R'_S{}^\beta{}_{\sigma\nu\mu} = ig^\beta{}_\sigma(F_S{}^1{}_{\nu\mu} + F_S{}^2{}_{\nu\mu})$$

$$R'_{\Omega}{}^{\beta}{}_{\sigma\nu\mu} = ig^{\beta}{}_{\sigma}(F_{\Omega}{}^{1}{}_{\nu\mu} + F_{\Omega}{}^{2}{}_{\nu\mu})$$

$$R'_{B}{}^{\beta}{}_{\sigma\nu\mu} = ig^{\beta}{}_{\sigma}(F_{B}{}^{1}{}_{\nu\mu} + F_{B}{}^{2}{}_{\nu\mu})$$

and

$$R'_{G}{}^{\beta}{}_{\sigma\nu\mu} = \partial_{\mu}H^{1\beta}{}_{\sigma\nu} - \partial_{\nu}H^{1\beta}{}_{\sigma\mu} + H^{1\gamma}{}_{\nu\sigma}H^{1\beta}{}_{\gamma\mu} - H^{1\gamma}{}_{\mu\sigma}H^{1\beta}{}_{\gamma\nu} + \partial_{\mu}H^{2\beta}{}_{\sigma\nu} - \partial_{\nu}H^{2\beta}{}_{\sigma\mu} +$$
$$+ H^{2\gamma}{}_{\nu\sigma}H^{2\beta}{}_{\gamma\mu} - H^{2\gamma}{}_{\mu\sigma}H^{2\beta}{}_{\gamma\nu} + H^{1\gamma}{}_{\nu\sigma}H^{2\beta}{}_{\gamma\mu} - H^{1\gamma}{}_{\mu\sigma}H^{2\beta}{}_{\gamma\nu} + H^{2\gamma}{}_{\nu\sigma}H^{1\beta}{}_{\gamma\mu} - \Gamma^{2\gamma}{}_{\mu\sigma}\Gamma^{\beta}{}_{\gamma\nu} \qquad (26.10)$$
$$= R^{1\beta}{}_{\sigma\nu\mu} + R^{2\beta}{}_{\sigma\nu\mu}$$

with

$$H^{\beta}{}_{\sigma\nu\mu} = \partial_{\mu}H^{\beta}{}_{\sigma\nu} - \partial_{\nu}H^{\beta}{}_{\sigma\mu} + H^{\gamma}{}_{\nu\sigma}H^{\beta}{}_{\gamma\mu} - H^{\gamma}{}_{\mu\sigma}H^{\beta}{}_{\gamma\nu} \qquad (26.11)$$

$$R^{1\beta}{}_{\sigma\nu\mu} = \partial_{\mu}H^{1\beta}{}_{\sigma\nu} - \partial_{\nu}H^{1\beta}{}_{\sigma\mu} + H^{1\gamma}{}_{\nu\sigma}H^{1\beta}{}_{\gamma\mu} - H^{1\gamma}{}_{\mu\sigma}H^{1\beta}{}_{\gamma\nu} \qquad (26.12)$$

$$R^{2\beta}{}_{\sigma\nu\mu p} = \partial_{\mu}H^{2\beta}{}_{\sigma\nu} - \partial_{\nu}H^{2\beta}{}_{\sigma\mu} + H^{2\gamma}{}_{\nu\sigma}H^{2\beta}{}_{\gamma\mu} - H^{2\gamma}{}_{\mu\sigma}H^{2\beta}{}_{\gamma\nu} +$$
$$+ H^{1\gamma}{}_{\nu\sigma}H^{2\beta}{}_{\gamma\mu} - H^{1\gamma}{}_{\mu\sigma}H^{2\beta}{}_{\gamma\nu} + H^{2\gamma}{}_{\nu\sigma}H^{1\beta}{}_{\gamma\mu} - H^{2\gamma}{}_{\mu\sigma}H^{1\beta}{}_{\gamma\nu} \qquad (26.13)$$

and

$$H^{1\sigma}{}_{\nu\mu} = \Gamma_{GR}{}^{\sigma}{}_{\nu\mu}$$

$$H^{2\sigma}{}_{\nu\mu} = \Gamma_{GR}{}^{2\sigma}{}_{\nu\mu}$$

and where

$$F_{SU(3)}{}^{1}{}_{\varkappa\mu} = \partial A_{SU(3)}{}^{1}{}_{\mu}/\partial x^{\varkappa} - \partial A_{SU(3)}{}^{1}{}_{\varkappa}/\partial x^{\mu} + ig_1[A_{SU(3)}{}^{1}{}_{\varkappa}, A_{U(3)}{}^{1}{}_{\mu}] \qquad (26.14)$$

$$F_{W}{}^{1}{}_{\varkappa\mu} = \partial W^{1}{}_{\mu}/\partial x^{\varkappa} - \partial W^{1}{}_{\varkappa}/\partial x^{\mu} + ig_2[W^{1}{}_{\varkappa}, W^{1}{}_{\mu}]$$

$$F_{E}{}^{1}{}_{\varkappa\mu} = \partial A_{E}{}^{1}{}_{\mu}/\partial x^{\varkappa} - \partial A_{E}{}^{1}{}_{\varkappa}/\partial x^{\mu}$$

$$F_{DW}{}^{1}{}_{\varkappa\mu} = \partial W_{D}{}^{1}{}_{\mu}/\partial x^{\varkappa} - \partial W_{D}{}^{1}{}_{\varkappa}/\partial x^{\mu} + ig_4[W_{D}{}^{1}{}_{\varkappa}, W_{D}{}^{1}{}_{\mu}]$$

$$F_{DE}{}^{1}{}_{\varkappa\mu} = \partial A_{DE}{}^{1}{}_{\mu}/\partial x^{\varkappa} - \partial A_{DE}{}^{1}{}_{\varkappa}/\partial x^{\mu}$$

$$F_{U}{}^{1}{}_{\varkappa\mu} = \partial U^{1}{}_{\mu}/\partial x^{\varkappa} - \partial U^{1}{}_{\varkappa}/\partial x^{\mu} + ig_6[U^{1}{}_{\varkappa}, U^{1}{}_{\mu}]$$

$$F_{V}{}^{1}{}_{\varkappa\mu} = \partial V^{1}{}_{\mu}/\partial x^{\varkappa} - \partial V^{1}{}_{\varkappa}/\partial x^{\mu} + ig_7[V^{1}{}_{\varkappa}, V^{1}{}_{\mu}]$$

$$F_{S}{}^{1}{}_{\varkappa\mu} = \partial A_{S}{}^{1}{}_{\mu}/\partial x^{\varkappa} - \partial A_{S}{}^{1}{}_{\varkappa}/\partial x^{\mu} + ig_8[A_{S}{}^{1}{}_{\varkappa}, A_{S}{}^{1}{}_{\mu}]$$

$$F_{\Omega}{}^{1}{}_{\varkappa\mu} = \partial A_{\Omega}{}^{1}{}_{\mu}/\partial x^{\varkappa} - \partial A_{\Omega}{}^{1}{}_{\varkappa}/\partial x^{\mu} + ig_{\Omega}[A_{\Omega}{}^{1}{}_{\varkappa}, A_{\Omega}{}^{1}{}_{\mu}]$$

$$F_{B}{}^{1}{}_{\varkappa\mu} = \partial B^{1}{}_{\mu}/\partial x^{\varkappa} - \partial B^{1}{}_{\varkappa}/\partial x^{\mu} + i[B^{1}{}_{\varkappa}, B^{1}{}_{\mu}]$$

$$F_{SU(3)}{}^{2}{}_{\varkappa\mu} = \partial A_{SU(3)}{}^{2}{}_{\mu}/\partial x^{\varkappa} - \partial A_{SU(3)}{}^{2}{}_{\varkappa}/\partial x^{\mu} + ig_1[A_{SU(3)}{}^{2}{}_{\varkappa}, A_{SU(3)}{}^{2}{}_{\mu}] + ig_1[A_{SU(3)}{}^{1}{}_{\varkappa}, A_{SU(3)}{}^{2}{}_{\mu}] +$$
$$+ ig_1[A_{SU(3)}{}^{2}{}_{\varkappa}, A_{SU(3)}{}^{1}{}_{\mu}]$$

$$F_{W}{}^{2}{}_{\varkappa\mu} = \partial W^{2}{}_{\mu}/\partial x^{\varkappa} - \partial W^{2}{}_{\varkappa}/\partial x^{\mu} + ig_2[W^{2}{}_{\varkappa}, W^{2}{}_{\mu}] + ig_2[W^{1}{}_{\varkappa}, W^{2}{}_{\mu}] + ig_2[W^{2}{}_{\varkappa}, W^{1}{}_{\mu}]$$

$$F_{E}{}^{2}{}_{\varkappa\mu} = \partial A_{E}{}^{2}{}_{\mu}/\partial x^{\varkappa} - \partial A_{E}{}^{2}{}_{\varkappa}/\partial x^{\mu}$$

$$F_{DW}{}^{2}{}_{\varkappa\mu} = \partial W_{D}{}^{2}{}_{\mu}/\partial x^{\varkappa} - \partial W_{D}{}^{2}{}_{\varkappa}/\partial x^{\mu} + ig_4[W_{D}{}^{2}{}_{\varkappa}, W_{D}{}^{2}{}_{\mu}] + ig_4[W_{D}{}^{1}{}_{\varkappa}, W_{D}{}^{2}{}_{\mu}] +$$

$$+ \, ig_4[W_{D\,\varkappa}^{2}, W_{D\,\mu}^{1}]$$

$$F_{DE\,\varkappa\mu}^{2} = \partial A_{DE\,\mu}^{2}/\partial x^{\varkappa} - \partial A_{DE\,\varkappa}^{2}/\partial x^{\mu}$$

$$F_{U\,\varkappa\mu}^{2} = \partial U_{\mu}^{2}/\partial x^{\varkappa} - \partial U_{\varkappa}^{2}/\partial x^{\mu} + ig_6[U_{\varkappa}^{2}, U_{\mu}^{2}] + ig_6[U_{\varkappa}^{1}, U_{\mu}^{2}] + ig_6[U_{\varkappa}^{2}, U_{\mu}^{1}]$$

$$F_{V\,\varkappa\mu}^{2} = \partial V_{\mu}^{2}/\partial x^{\varkappa} - \partial V_{\varkappa}^{2}/\partial x^{\mu} + ig_7[V_{\varkappa}^{2}, V_{\mu}^{2}] + ig_7[V_{\varkappa}^{1}, V_{\mu}^{2}] + ig_7[V_{\varkappa}^{2}, V_{\mu}^{1}]$$

$$F_{S\,\varkappa\mu}^{2} = \partial A_{S\,\mu}^{2}/\partial x^{\varkappa} - \partial A_{S\,\varkappa}^{2}/\partial x^{\mu} + ig_8[A_{S\,\varkappa}^{2}, A_{S\,\mu}^{2}] + ig_8[A_{S\,\varkappa}^{1}, A_{S\,\mu}^{2}] +$$
$$+ \, ig_8[A_{S\,\varkappa}^{2}, A_{S\,\mu}^{1}]$$

$$F_{\Omega\,\varkappa\mu}^{2} = \partial A_{\Omega\,\mu}^{2}/\partial x^{\varkappa} - \partial A_{\Omega\,\varkappa}^{2}/\partial x^{\mu} + ig_{\Omega}[A_{\Omega\,\varkappa}^{2}, A_{\Omega\,\mu}^{2}] + ig_{\Omega}[A_{\Omega\,\varkappa}^{1}, A_{\Omega\,\mu}^{2}] + ig_{\Omega}[A_{\Omega\,\varkappa}^{2}, A_{\Omega\,\mu}^{1}]$$

$$F_{B\,\varkappa\mu}^{2} = \partial B_{\mu}^{2}/\partial x^{\varkappa} - \partial B_{\varkappa}^{2}/\partial x^{\mu} + i[B_{\mu}^{2}, B_{\varkappa}^{2}] + i[B_{\mu}^{1}, B_{\varkappa}^{2}] + i[B_{\mu}^{2}, B_{\varkappa}^{1}]$$

Note that $R'^{\beta}_{\sigma\nu\mu}$ factorizes into $U(1)\otimes SU(2)\otimes U(1)\otimes SU(2)\otimes SU(3)\otimes U(4)\otimes U(4)\otimes U(4)\otimes U(4)\otimes U(64)$ parts and a Riemann-Christoffel Gravitational curvature tensor part. For later use in defining a lagrangian we define

$$R'^{\beta}_{\sigma\nu\mu} = R'^{1\beta}_{E\,\sigma\nu\mu} + R'^{2\beta}_{E\,\sigma\nu\mu} + R'^{1\beta}_{SU(2)\,\sigma\nu\mu} + R'^{2\beta}_{SU(2)\,\sigma\nu\mu} + R'^{1\beta}_{DE\,\sigma\nu\mu} + R'^{2\beta}_{DE\,\sigma\nu\mu} + R'^{1\beta}_{DSU(2)\,\sigma\nu\mu} +$$
$$+ \, R'^{2\beta}_{DSU(2)\,\sigma\nu\mu} + R'^{1\beta}_{SU(3)\,\sigma\nu\mu} + R'^{2\beta}_{SU(3)\,\sigma\nu\mu} + R'^{1\beta}_{U\,\sigma\nu\mu} + R'^{2\beta}_{U\,\sigma\nu\mu} + R'^{1\beta}_{V\,\sigma\nu\mu} + R'^{2\beta}_{V\,\sigma\nu\mu} +$$
$$+ \, R'^{1\beta}_{S\,\sigma\nu\mu} + R'^{2\beta}_{S\,\sigma\nu\mu} + R'^{1\beta}_{\Omega\,\sigma\nu\mu} + R'^{2\beta}_{\Omega\,\sigma\nu\mu} + R'^{1\beta}_{B\,\sigma\nu\mu} + R'^{2\beta}_{B\,\sigma\nu\mu} + R^{1\beta}_{\sigma\nu\mu} + R^{2\beta}_{\sigma\nu\mu}$$

$$(26.15)$$

where

$$R'^{1\beta}_{E\,\sigma\nu\mu} = ig^{\beta}_{\sigma}F_{E\,\nu\mu}^{1}$$
$$R'^{2\beta}_{E\,\sigma\nu\mu} = ig^{\beta}_{\sigma}F_{DE\,\nu\mu}^{2}$$

$$R'^{1\beta}_{DE\,\sigma\nu\mu} = ig^{\beta}_{\sigma}F_{E\,\nu\mu}^{1}$$
$$R'^{2\beta}_{DE\,\sigma\nu\mu} = ig^{\beta}_{\sigma}F_{DE\,\nu\mu}^{2}$$

$$R'^{1\beta}_{SU(2)\,\sigma\nu\mu} = ig^{\beta}_{\sigma}F_{W\,\nu\mu}^{1}$$
$$R'^{2\beta}_{SU(2)\,\sigma\nu\mu} = ig^{\beta}_{\sigma}F_{DW\,\nu\mu}^{2}$$

$$R'^{1\beta}_{DSU(2)\,\sigma\nu\mu} = ig^{\beta}_{\sigma}F_{W\,\nu\mu}^{1}$$
$$R'^{2\beta}_{DSU(2)\,\sigma\nu\mu} = ig^{\beta}_{\sigma}F_{DW\,\nu\mu}^{2}$$

$$R'^{1\beta}_{SU(3)\,\sigma\nu\mu} = ig^{\beta}_{\sigma}F_{SU(3)\,\nu\mu}^{1}$$
$$R'^{2\beta}_{SU(3)\,\sigma\nu\mu} = ig^{\beta}_{\sigma}F_{SU(3)\,\nu\mu}^{2}$$

$$R'^{1\beta}_{U\,\sigma\nu\mu} = ig^{\beta}_{\sigma}F_{U\,\nu\mu}^{1}$$
$$R'^{2\beta}_{U\,\sigma\nu\mu} = ig^{\beta}_{\sigma}F_{U\,\nu\mu}^{2}$$

$$R'_V{}^{1\beta}{}_{\sigma\nu\mu} = ig^\beta{}_\sigma F_V{}^1{}_{\nu\mu}$$
$$R'_V{}^{2\beta}{}_{\sigma\nu\mu} = ig^\beta{}_\sigma F_V{}^2{}_{\nu\mu}$$

$$R'_S{}^{1\beta}{}_{\sigma\nu\mu} = ig^\beta{}_\sigma F_S{}^1{}_{\nu\mu}$$
$$R'_S{}^{2\beta}{}_{\sigma\nu\mu} = ig^\beta{}_\sigma F_S{}^2{}_{\nu\mu}$$
$$R'_\Omega{}^{1\beta}{}_{\sigma\nu\mu} = ig^\beta{}_\sigma F_\Omega{}^1{}_{\nu\mu}$$
$$R'_\Omega{}^{2\beta}{}_{\sigma\nu\mu} = ig^\beta{}_\sigma F_\Omega{}^2{}_{\nu\mu}$$

$$R'_B{}^{1\beta}{}_{\sigma\nu\mu} = ig^\beta{}_\sigma B^1{}_{\nu\mu}$$
$$R'_B{}^{2\beta}{}_{\sigma\nu\mu} = ig^\beta{}_\sigma B^2{}_{\nu\mu}$$

The total Ricci tensor is

$$R'_{\sigma\mu} = R'^{\beta}{}_{\sigma\beta\mu} \tag{26.16}$$

$$= iF_E{}^1{}_{\sigma\mu} + iF_E{}^2{}_{\sigma\mu} + iF_W{}^1{}_{\sigma\mu} + iF_W{}^2{}_{\sigma\mu} + iF_{DE}{}^1{}_{\sigma\mu} + iF_{DE}{}^2{}_{\sigma\mu} + iF_{DW}{}^1{}_{\sigma\mu} + iF_{DW}{}^2{}_{\sigma\mu} + iF_{SU(3)}{}^1{}_{\sigma\mu} + iF_{SU(3)}{}^2{}_{\sigma\mu} +$$
$$+ iF_U{}^1{}_{\sigma\mu} + iF_U{}^2{}_{\sigma\mu} + iF_V{}^1{}_{\sigma\mu} + iF_V{}^2{}_{\sigma\mu} + iF_S{}^1{}_{\sigma\mu} + iF_S{}^2{}_{\sigma\mu} + iF_\Omega{}^1{}_{\sigma\mu} + iF_\Omega{}^2{}_{\sigma\mu} + iB^1{}_{\sigma\mu} + iB^2{}_{\sigma\mu} +$$
$$+ \partial_\mu H^{1\beta}{}_{\sigma\beta} - \partial_\beta H^{1\beta}{}_{\sigma\mu} + H^{1\gamma}{}_{\beta\sigma} H^{1\beta}{}_{\gamma\mu} - H^{1\gamma}{}_{\mu\sigma} H^{1\beta}{}_{\gamma\beta} +$$
$$+ \partial_\mu H^{2\beta}{}_{\sigma\beta} - \partial_\beta H^{2\beta}{}_{\sigma\mu} + H^{2\gamma}{}_{\beta\sigma} H^{2\beta}{}_{\gamma\mu} - H^{2\gamma}{}_{\mu\sigma} H^{2\beta}{}_{\gamma\beta} + H^{1\gamma}{}_{\beta\sigma} H^{2\beta}{}_{\gamma\mu} - H^{1\gamma}{}_{\mu\sigma} H^{2\beta}{}_{\gamma\beta} + H^{2\gamma}{}_{\beta\sigma} H^{1\beta}{}_{\gamma\mu} - H^{2\gamma}{}_{\mu\sigma} H^{1\beta}{}_{\gamma\beta}$$

$$= R'_E{}^1{}_{\sigma\mu} + R'_E{}^2{}_{\sigma\mu} + R'_{SU(2)}{}^1{}_{\sigma\mu} + R'_{SU(2)}{}^2{}_{\sigma\mu} + R'_{DE}{}^1{}_{\sigma\mu} + R'_{DE}{}^2{}_{\sigma\mu} + R'_{DSU(2)}{}^1{}_{\sigma\mu} + R'_{DSU(2)}{}^2{}_{\sigma\mu} + R'_{SU(3)}{}^1{}_{\sigma\mu} +$$
$$+ R'_{SU(3)}{}^2{}_{\sigma\mu} + R'_U{}^1{}_{\sigma\mu} + R'_U{}^2{}_{\sigma\mu} + R'_V{}^1{}_{\sigma\mu} + R'_V{}^2{}_{\sigma\mu} + R'_S{}^1{}_{\sigma\mu} + R'_S{}^2{}_{\sigma\mu} + R'_\Omega{}^1{}_{\sigma\mu} + R'_\Omega{}^2{}_{\sigma\mu} + R'_B{}^{1\beta}{}_{\sigma\beta\mu} +$$
$$+ R'_B{}^{2\beta}{}_{\sigma\beta\mu} + R^1{}_{\sigma\mu} + R^2{}_{\sigma\mu}$$
$$= R'^1{}_{\sigma\mu} + R'^2{}_{\sigma\mu} \tag{26.17}$$

where

$$R'^1{}_{\sigma\mu} = R'_E{}^1{}_{\sigma\mu} + R'_{SU(2)}{}^1{}_{\sigma\mu} + R'_{DE}{}^1{}_{\sigma\mu} + R'_{DSU(2)}{}^1{}_{\sigma\mu} + R'_{SU(3)}{}^1{}_{\sigma\mu} + R'_U{}^1{}_{\sigma\mu} + R'_V{}^1{}_{\sigma\mu} + R'_S{}^1{}_{\sigma\mu} +$$
$$+ R'_\Omega{}^1{}_{\sigma\mu} + R'_B{}^{1\beta}{}_{\sigma\beta\mu} + R^1{}_{\sigma\mu} \tag{26.18}$$

$$R'^2{}_{\sigma\mu} = R'_E{}^2{}_{\sigma\mu} + R'_{SU(2)}{}^2{}_{\sigma\mu} + R'_{DE}{}^2{}_{\sigma\mu} + R'_{DSU(2)}{}^2{}_{\sigma\mu} + R'_{SU(3)}{}^2{}_{\sigma\mu} + R'_U{}^2{}_{\sigma\mu} + R'_V{}^2{}_{\sigma\mu} + R'_S{}^2{}_{\sigma\mu} +$$
$$+ R'_\Omega{}^2{}_{\sigma\mu} + R'_B{}^{2\beta}{}_{\sigma\beta\mu} + R^2{}_{\sigma\mu}$$

with

$$R'_E{}^1{}_{\sigma\mu} = iF_E{}^1{}_{\sigma\mu}$$
$$R'_E{}^2{}_{\sigma\mu} = iF_E{}^2{}_{\sigma\mu}$$

$$R'^{\;1}_{SU(2)\;\sigma\mu} = iF^{\;1}_{W\;\sigma\mu}$$
$$R'^{\;2}_{SU(2)\;\sigma\mu} = iF^{\;2}_{W\;\sigma\mu}$$

$$R'^{\;1}_{DE\;\sigma\mu} = iF^{\;1}_{DE\;\sigma\mu}$$
$$R'^{\;2}_{DE\;\sigma\mu} = iF^{\;2}_{DE\;\sigma\mu}$$

$$R'^{\;1}_{DSU(2)\;\sigma\mu} = iF^{\;1}_{DW\;\sigma\mu}$$
$$R'^{\;2}_{DSU(2)\;\sigma\mu} = iF^{\;2}_{DW\;\sigma\mu}$$

$$R'^{\;1}_{SU(3)\;\sigma\mu} = iF^{\;1}_{SU(3)\;\sigma\mu}$$
$$R'^{\;2}_{SU(3)\;\sigma\mu} = iF^{\;2}_{SU(3)\;\sigma\mu}$$

$$R'^{\;1}_{U\;\sigma\mu} = iF^{\;1}_{U\;\sigma\mu}$$
$$R'^{\;2}_{U\;\sigma\mu} = iF^{\;2}_{U\;\sigma\mu}$$

$$R'^{\;1}_{V\;\sigma\mu} = iF^{\;1}_{V\;\sigma\mu}$$
$$R'^{\;2}_{V\;\sigma\mu} = iF^{\;2}_{V\;\sigma\mu}$$

$$R'^{\;1}_{S\;\sigma\mu} = iF^{\;1}_{S\;\sigma\mu}$$
$$R'^{\;2}_{S\;\sigma\mu} = iF^{\;2}_{S\;\sigma\mu}$$

$$R'^{\;1}_{\Omega\;\sigma\mu} = iF^{\;1}_{\Omega\;\sigma\mu}$$
$$R'^{\;2}_{\Omega\;\sigma\mu} = iF^{\;2}_{\Omega\;\sigma\mu}$$

$$R'^{\;1}_{B\;\sigma\mu} = iB^{\;1}_{\sigma\mu}$$
$$R'^{\;2}_{B\;\sigma\mu} = iB^{\;2}_{\sigma\mu}$$

with the further definition of $R''^{\;1}_{\sigma\mu}$ and $R''^{\;2}_{\sigma\mu}$:

$$R''^{\;1}_{\sigma\mu} = R'^{\;1}_{SU(3)\;\sigma\mu} + R^{\;1}_{\sigma\mu} \qquad (26.19)$$
$$R''^{\;2}_{\sigma\mu} = R'^{\;2}_{SU(3)\;\sigma\mu} + R^{\;2}_{\sigma\mu}$$

Eq. 26.18 is the Ricci tensor. An additional Ricci-like tensor is

$$H_{\sigma\mu} = H^{\beta}_{\;\sigma\beta\mu} \qquad (26.20)$$

The curvature scalar is

$$R' = g^{\sigma\mu}R'_{\sigma\mu} = + \partial^\sigma H^{1\beta}{}_{\sigma\beta} - \partial_\beta H^{1\beta}{}_\sigma{}^\sigma + H^{1\gamma}{}_{\beta\sigma}H^{1\beta}{}_\gamma{}^\sigma - H^{1\gamma}{}_{\mu\sigma}H^{1\beta}{}_{\gamma\beta} + \partial^\sigma H^{2\beta}{}_{\sigma\beta} - \partial_\beta H^{2\beta}{}_\sigma{}^\sigma +$$
$$+ H^{2\gamma}{}_{\beta\sigma}H^{2\beta}{}_\gamma{}^\sigma - H^{2\gamma\sigma}{}_\sigma H^{2\beta}{}_{\gamma\beta} + H^{1\gamma}{}_{\beta\sigma}H^{2\beta}{}_\gamma{}^\sigma - H^{1\gamma\sigma}{}_\sigma H^{2\beta}{}_{\gamma\beta} + H^{2\gamma}{}_{\beta\sigma}H^{1\beta}{}_\gamma{}^\sigma - H^{2\gamma\sigma}{}_\sigma H^{1\beta}{}_{\gamma\beta}$$

$$= g^{\sigma\mu}(R^{1\beta}{}_{\sigma\beta\mu} + R^{2\beta}{}_{\sigma\beta\mu}) \qquad\qquad (26.21)$$

1.6 Lagrangian Terms

The boson part of the lagrangian of the unified Extended Standard Model (with the Higgs sector and the Faddeev-Popov terms gauge sector not displayed here) is:

$$\mathcal{L} = \mathrm{Tr}\ \sqrt{g}[MD_v R''^1{}_{\sigma\mu}D^v R''^{2\sigma\mu} + aR'^1{}_{\sigma\mu}R'^{2\sigma\mu} + bR' + cg^{\sigma\mu}g^2{}_{\sigma\mu} + c'g^{2\sigma\mu}g^2{}_{\sigma\mu} - dA_{SU(3)}{}^2{}_\mu A_{SU(3)}{}^{2\mu}] \tag{27.1}$$

where M, a, b, c, c', and d are constants, and $R''^i{}_{\sigma\mu}$ for i = 1, 2 determined by eqs. 26.12 and 26.13.[18]

This higher derivative lagrangian maintains the locality of the theory but does entail a modest modification in the derivation of the Euler-Lagrange equations of motion. It also requires the use of principal value propagators rather than ordinary Feynman propagators for gluon and graviton interactions. Thus the Strong Interaction sector, and the Gravitation sector are Action-at-a-Distance theories that are similar in spirit to Wheeler-Feynman Electrodynamics. The two U(1) Electromagnetic sectors, the Generation group U(4) gauge field sector, the Layer group U(4) gauge field sector, the two SU2) Weak sectors, the U(4) A_s gauge field sector, the spinor connection sector, and the Ω-interaction sector may, or may not, be

[18] One may ask why $R''^1{}_{\sigma\mu}$ and $R''^2{}_{\sigma\mu}$ appear in the first term of the lagrangian, and not other interaction terms. We believe the primary reason is: "The extended vierbein $l^{\mu ai}(x)$ can be viewed as located at a point in a higher dimensional complex-valued space.

$$l^{\mu ai}(x) = (\partial \xi_X{}^{ai}(x)/\partial x_\mu)_{X=h(x)}$$

where $\xi_X{}^{ai}$ is a set of locally inertial coordinates located at point X, and x = h(x) is a 4-dimensional point in a tangent subspace of the higher dimensional space:

$$X = h(x)$$

The relation between complex 4-dimensional coordinates x and the higher dimensional coordinates X is an embedding of a 4-dimensional surface within the higher dimensional complex space when account is taken of the range of possible x values. We have considered such embeddings in Blaha (2015a), and in earlier books, and developed a theory of a higher dimensional complex-valued space (the *Megaverse*) that contains our universe and probably many other universes." Thus SU(3) and Gravitation have a special role in our particle dynamics based on geometry. The second reason is the common feature of color SU(3) and real-valued General Relativity is that they are the only interactions that do not participate in 'rotations of interactions' as described earlier and in chapter 31 of volume I. The third, practical reason is the experimental reality that the Strong Interaction and Gravitation are known to have 'anomalous' features that will be seen to be remedied by these insertions while the other interactions are 'conventional.'

Action-at-a-Distance fields. They are not constrained to be Action-at-a-Distance by the present considerations.

Since we wish to apply our theory cosmologically, and within hadrons, where the gravitational spinor connections are negligible due to the smallness of the gravitational constant G and the 'smallness' of B spin on the cosmological scale, we set $B^1_{\nu\mu} = B^2_{\nu\mu} = 0$ and find[19]

$$\mathcal{L} = \text{Tr } \sqrt{g}[MD_\nu(R''^1{}_{SU(3)\sigma\mu} + R'_G{}^1{}_{\sigma\mu})D^\nu(R'_{SU(3)}{}^{2\sigma\mu} + R'_G{}^{2\sigma\mu}) +$$
$$+ aR''^1{}_{\sigma\mu}R'^{2\sigma\mu} + bR' + cg^{\sigma\mu}g^2{}_{\sigma\mu} + c'g^{2\sigma\mu}g^2{}_{\sigma\mu} - dA_{SU(3)}{}^2{}_\mu A_{SU(3)}{}^{2\mu}] \qquad (27.2)$$

Since there are no strong interaction fields in 'empty' space and gravity is negligible within hadrons,[20] we can drop the interaction terms between the Strong interaction and the Gravity interaction. However, we cannot drop the interaction terms amongst Electromagnetism, the Weak interaction, the Strong Interaction, the Generation group U(4) interaction, the Layer group U(4) interaction, the U(4) Generaly Relativity Reality group interaction, and the SU(3)⊗U(64) Ω-interaction – within, and between, hadrons. The interaction terms between Electromagnetism and Gravitation are important cosmologically.

Eq. 27.2 can therefore be expressed as:[21]

$$\mathcal{L} = \mathcal{L}_E + \mathcal{L}_{SU(2)} + \mathcal{L}_{DE} + \mathcal{L}_{DSU(2)} + \mathcal{L}_{SU(3)} + \mathcal{L}_U + \mathcal{L}_V + \mathcal{L}_S + \mathcal{L}_\Omega + \mathcal{L}_G + \mathcal{L}_{int} \qquad (27.3)$$

where taking traces of $\mathcal{L}$s terms is understood

$$\mathcal{L}_E = \text{Tr } \sqrt{g}\{M\{[\partial_\nu + i(A_E{}^1{}_\nu + A_E{}^2{}_\nu)]F^1_{E\sigma\mu}[\partial^\nu + i(A_E{}^{1\nu} + A_E{}^{2\nu})]F^2_E{}^{\sigma\mu}\} + aF_E{}^1{}_{\sigma\mu}F_E{}^{2\sigma\mu}\}$$

$$(27.4)$$

$$\mathcal{L}_{SU(2)} = \text{Tr } \sqrt{g}[aF_W{}^1{}_{\sigma\mu}F_W{}^{2\sigma\mu}]$$
$$\mathcal{L}_{DE} = \text{Tr } \sqrt{g}\{M\{[\partial_\nu + i(A_{DE}{}^1{}_\nu + A_{DE}{}^2{}_\nu)]F^1_{DE\sigma\mu}[\partial^\nu + i(A_{DE}{}^{1\nu} + A_{DE}{}^{2\nu})]F_{DE}{}^{2\sigma\mu}\} + aF_{DE}{}^1{}_{\sigma\mu}F_{DE}{}^{2\sigma\mu}\}$$
$$\mathcal{L}_{DSU(2)} = \text{Tr } \sqrt{g}[aF_W{}^1{}_{\sigma\mu}F_W{}^{2\sigma\mu}]$$
$$\mathcal{L}_{SU(3)} = \text{Tr } \sqrt{g}\{M[\partial_\nu + i(A_{SU(3)}{}^1{}_\nu + A_{SU(3)}{}^2{}_\nu)]F_{SU(3)}{}^1{}_{\sigma\mu}[\partial^\nu + i(A_{SU(3)}{}^{1\nu} + A_{SU(3)}{}^{2\nu})]F_{SU(3)}{}^{2\sigma\mu} +$$
$$+ aF_{SU(3)}{}^1{}_{\sigma\mu}F_{SU(3)}{}^{2\sigma\mu} - dA_{SU(3)}{}^2{}_\mu A_{SU(3)}{}^{2\mu}\}$$
$$\mathcal{L}_U = \text{Tr } \sqrt{g}[aF_U{}^1{}_{\sigma\mu}F_U{}^{2\sigma\mu}]$$
$$\mathcal{L}_V = \text{Tr } \sqrt{g}[aF_V{}^1{}_{\sigma\mu}F_V{}^{2\sigma\mu}]$$
$$\mathcal{L}_S = \text{Tr } \sqrt{g}[aF_S{}^1{}_{\sigma\mu}F_S{}^{2\sigma\mu}]$$
$$\mathcal{L}_\Omega = \text{Tr } \sqrt{g}[aF_\Omega{}^1{}_{\sigma\mu}F_\Omega{}^{2\sigma\mu}]$$

[19] The constants have the dimensions: M has the dimension of inverse mass squared, b has dimension mass squared, a is dimensionless, c and c' have dimension mass, and d has dimension mass squared.

[20] We show gravity weakens at very short distances using our Two-Tier Quantum Field Theory formalism. See Appendix A, and Blaha (2003) and (2005a) among other books by the author.

[21] We only consider the gauge field lagrangian terms.

$$\mathcal{L}_G = \mathrm{Tr}\ \sqrt{g}[MD_v R^1{}_{\sigma\mu}D^v R^{2\sigma\mu} + aR^1{}_{\sigma\mu}R^{2\sigma\mu} + bg^{\sigma\mu}(R^{1\beta}{}_{\sigma\beta\mu} + R^{2\beta}{}_{\sigma\beta\mu}) + cg^{\sigma\mu}g^2{}_{\sigma\mu} + c'g^{2\sigma\mu}g^2{}_{\sigma\mu}]$$
$$= \mathrm{Tr}\ \sqrt{g}[MD_v R^1{}_{\sigma\mu}D^v R^{2\sigma\mu} + aR^1{}_{\sigma\mu}R^{2\sigma\mu} + bH + cg^{\sigma\mu}g^2{}_{\sigma\mu} + c'g^{2\sigma\mu}g^2{}_{\sigma\mu}]$$

$$\mathcal{L}_{int} = \mathcal{L} - (\mathcal{L}_E + \mathcal{L}_{SU(2)} + \mathcal{L}_{DE} + \mathcal{L}_{DSU(2)} + \mathcal{L}_{SU(3)} + \mathcal{L}_U + \mathcal{L}_V + \mathcal{L}_S + \mathcal{L}_\Omega + \mathcal{L}_G)$$

Thus $\mathcal{L}_{SU(3)}$, $\mathcal{L}_{SU(2)}$, $\mathcal{L}_E$, $\mathcal{L}_{DE}$, $\mathcal{L}_{DSU(2)}$, $\mathcal{L}_U$, $\mathcal{L}_V$, $\mathcal{L}_S$, $\mathcal{L}_\Omega$, and parts of $\mathcal{L}_{int}$ are the dominant interactions within hadrons, and $\mathcal{L}_G$, $\mathcal{L}_E$ and parts of $\mathcal{L}_{int}$ are the dominant interactions in space within the framework of this discussion.

The $D_v R^1{}_{\sigma\mu}$ and $D^v R^{2\sigma\mu}$ terms have the form:

$$D_v R^i{}_{\sigma\mu} = + \partial_v R^i{}_{\sigma\mu} - H^{1\beta}{}_{\sigma v}R^i{}_{\beta\mu} - H^{1\beta}{}_{v\mu}R^i{}_{\sigma\beta}$$

for i = 1, 2.

1.7 New Vector Boson Interactions

The above lagrangian can be broken up into pieces in the following manner:

$$\mathcal{L}_E = \mathrm{Tr}\ \sqrt{g}\{M\{[\partial_v + i(A_E^1{}_v + A_E^2{}_v)]F^1{}_{E\sigma\mu}[\partial^v + i(A_E^{1v} + A_E^{2v})]F^2{}_E{}^{\sigma\mu}\} + aF_E^1{}_{\sigma\mu}F_E^{2\sigma\mu}\} \qquad (27.4)$$
$$\mathcal{L}_{SU(2)} = \mathrm{Tr}\ \sqrt{g}[aF_W^1{}_{\sigma\mu}F_W^{2\sigma\mu}]$$
$$\mathcal{L}_{DE} = \mathrm{Tr}\ \sqrt{g}\{M\{[\partial_v + i(A_{DE}^1{}_v + A_{DE}^2{}_v)]F^1{}_{DE\sigma\mu}[\partial^v + i(A_{DE}^{1v} + A_{DE}^{2v})]F_{DE}^{2\sigma\mu}\} + aF_{DE}^1{}_{\sigma\mu}F_{DE}^{2\sigma\mu}\}$$
$$\mathcal{L}_{DSU(2)} = \mathrm{Tr}\ \sqrt{g}[aF_W^1{}_{\sigma\mu}F_W^{2\sigma\mu}]$$
$$\mathcal{L}_{SU(3)} = \mathrm{Tr}\ \sqrt{g}\{M[\partial_v + i(A_{SU(3)}^1{}_v + A_{SU(3)}^2{}_v)]F_{SU(3)}^1{}_{\sigma\mu}[\partial^v + i(A_{SU(3)}^{1v} + A_{SU(3)}^{2v})]F_{SU(3)}^{2\sigma\mu} +$$
$$+ aF_{SU(3)}^1{}_{\sigma\mu}F_{SU(3)}^{2\sigma\mu} - dA_{SU(3)}^2{}_\mu A_{SU(3)}^{2\mu}\}$$
$$\mathcal{L}_U = \mathrm{Tr}\ \sqrt{g}[aF_U^1{}_{\sigma\mu}F_U^{2\sigma\mu}]$$
$$\mathcal{L}_V = \mathrm{Tr}\ \sqrt{g}[aF_V^1{}_{\sigma\mu}F_V^{2\sigma\mu}]$$
$$\mathcal{L}_S = \mathrm{Tr}\ \sqrt{g}[aF_S^1{}_{\sigma\mu}F_S^{2\sigma\mu}]$$
$$\mathcal{L}_\Omega = \mathrm{Tr}\ \sqrt{g}[aF_\Omega^1{}_{\sigma\mu}F_\Omega^{2\sigma\mu}]$$
$$\mathcal{L}_G = \mathrm{Tr}\ \sqrt{g}[MD_v R^1{}_{\sigma\mu}D^v R^{2\sigma\mu} + aR^1{}_{\sigma\mu}R^{2\sigma\mu} + bg^{\sigma\mu}(R^{1\beta}{}_{\sigma\beta\mu} + R^{2\beta}{}_{\sigma\beta\mu}) + cg^{\sigma\mu}g^2{}_{\sigma\mu} + c'g^{2\sigma\mu}g^2{}_{\sigma\mu}]$$
$$= \mathrm{Tr}\ \sqrt{g}[MD_v R^1{}_{\sigma\mu}D^v R^{2\sigma\mu} + aR^1{}_{\sigma\mu}R^{2\sigma\mu} + bH + cg^{\sigma\mu}g^2{}_{\sigma\mu} + c'g^{2\sigma\mu}g^2{}_{\sigma\mu}]$$

$$\mathcal{L}_{int} = \mathcal{L} - (\mathcal{L}_E + \mathcal{L}_{SU(2)} + \mathcal{L}_{DE} + \mathcal{L}_{DSU(2)} + \mathcal{L}_{SU(3)} + \mathcal{L}_U + \mathcal{L}_V + \mathcal{L}_S + \mathcal{L}_\Omega + \mathcal{L}_G)$$

Thus $\mathcal{L}_{SU(3)}$, $\mathcal{L}_{SU(2)}$, $\mathcal{L}_E$, $\mathcal{L}_{DE}$, $\mathcal{L}_{DSU(2)}$, $\mathcal{L}_U$, $\mathcal{L}_V$, $\mathcal{L}_S$, $\mathcal{L}_\Omega$, and parts of $\mathcal{L}_{int}$ are the dominant interactions within hadrons, and $\mathcal{L}_G$, $\mathcal{L}_E$ and parts of $\mathcal{L}_{int}$ are the dominant interactions in space within the framework of this discussion. The terms of $\mathcal{L}_{int}$ have 'new' interactions between gauge fields that are described in some detail in volume I. These interactions are not in the conventional Standard Model. They lead to modifications of gravity, the Strong Interactions, spin dynamics and so on.

2. A Unified SuperStandard Model

In the Extended Standard Model that we presented in volume I and earlier books we introduced the concept of a Layer group with four layers of fermions. We did not enlarge the $SU(3) \otimes SU(2) \otimes U(1) \otimes SU(2) \otimes U(1) \otimes U(4)$ vector bosons[22] to four layers. Consequently the Extended Standard Model would support interactions between fermions in different layers unless the masses of fermions in the higher three layers were much larger than current accelerator energies or the Layer group coupling constant is extraordinarily small. Although larger higher layer fermion masses would resolve the issue of interactions and transitions between fermion layers (which have not been observed), it seems that having different sets of $SU(3) \otimes SU(2) \otimes U(1) \otimes SU(2) \otimes U(1) \otimes U(4)$ vector bosons for each layer is a much more robust solution.

In addition the 'rotations of interactions' group and its vector bosons must be changed to reflect the new addition of layers of vector bosons. We therefore add these changes to the Extended Standard Model and call the new theory the Unified SuperStandard Model because it has a form of SuperSymmetry that we will describe later.

2.1 A New Form of the Layer Group

This section[23] describes a new form of the Layer group. The Complex Lorentz group transformations led to the Reality group: $R = SU(3) \otimes SU(2) \otimes U(1) \otimes SU(2) \otimes U(1)$, which is the symmetry group of the 'original' Standard Model. The U(4) Generation group was shown to follow from the four particle number conservation laws,[24] and increased the Standard Model group to $SU(3) \otimes SU(2) \otimes U(1) \otimes SU(2) \otimes U(1) \otimes U(4)$.

We will now show that the four layers of each of the four fermion generations also have (broken) particle number conservation laws and can embody a broken U(4) symmetry *for each generation* resulting in a $U(4)^4$ symmetry. (Fig. 2.1 shows the four layers of four generations.)

The rationale for the new Layer groups is similar to that of the Generation group which was based on the four particle numbers B, L, B_D, and L_D. The Generation group was based on U(4) rotations of the four number operators B, L, B_D, and L_D in the one generation Extended Standard Model. We can visualize these rotations horizontally as

[22] The U(4) interaction in this product is the Generation group interaction.
[23] Much of this chapter appears in an earlier form in Blaha (2016a) and (2016b).
[24] Baryon number, Lepton number, and their Dark analogues.

Figure 2.1. The four layers of fermions. Each layer (oval) has four generations of fermions. Layer 1 is the layer that we have found experimentally. The 4[th] generation of layer 1, the Dark part of layer 1, and the remaining three more massive layers constitute Dark matter.

After generations are introduced, then it becomes possible to consider *vertical* rotations amongst the four layers for each generation. For example the vertical U(4) rotations for generation 1 are symbolized by:

$$G_1$$
$$G_1$$
$$G_1$$
$$G_1$$

Rotations are based on four 'conserved' numbers L_1, L_2, L_3, and L_4 that sum the total number of fermions over the four generations in all layers within a state.[25] *L_i counts the number of fundamental fermions in generation i summed over all the layers for i = 1, 2, 3, 4.*

[25] The generations are numbered from 1 to 4 with the lowest masses generation (e, v_e, u, d), and its Dark analogues, being generation 1.

THE FERMION PERIODIC TABLE

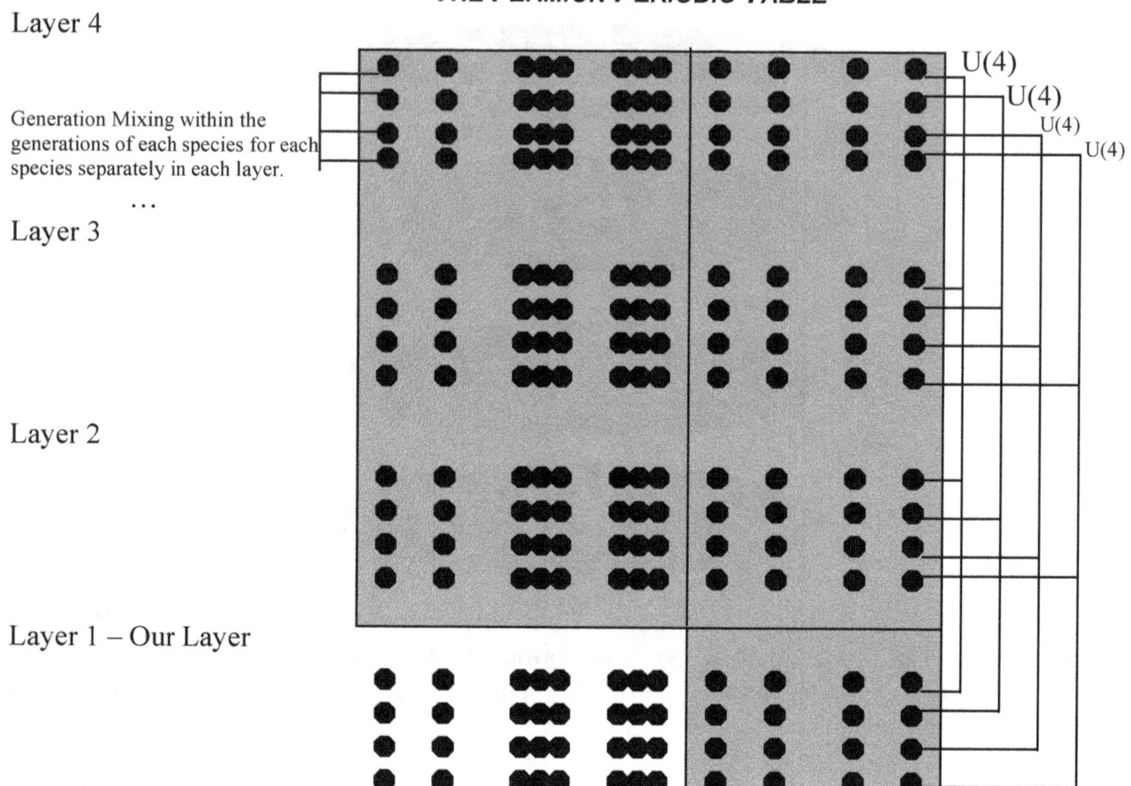

Figure 2.2. Partial example of pattern of mass mixing of the Generation group and of the Layer groups.. Dark parts of the periodic table are gray. Light parts are the known fermions with an additional, as yet not found, 4th generation shown. The lines on the left side show an example of the Generation mixing within one species. The Generation mixing applies to each species in each layer. The lines on the right side show the Layer mixing generation by generation.among all four layers for each generation individually.

Fermions have positive $L_i = +1$ values and anti-fermions have negative $L_i = -1$ values. For example, if a state has 3 u quarks, 1 d quark, 1 anti-s quark, 2 electrons, 2 anti-τ leptons, 2 Dark muon neutrinos, and one Dark electron neutrino ν_{De} then $L_1 = 3+1+2+1 = 7$, $L_2 = -1+2 = 1$, $L_3 = -2$ and $L_4 = 0$.

2.1.1 Four Conserved Layer Numbers per Generation

The total layer number L_i per generation i is the sum of four other 'conserved' layer numbers:

$$L_i = L_{iB} + L_{iB} + L_{iDB} + L_{iL} + L_{iDL} \qquad (2.1)$$

where

L_{iB}　– The Bayon layer number
L_{iDB}　– The Dark Baryon layer number
L_{iL}　– The Lepton layer number
L_{iDL}　– The Dark Lepton layer number

Thus each generation of the four layers has four 'conserved' layer particle numbers like the Generation group. Since we can perform complex rotations among the four layers of generation I and since there are four diagonal generators they must each be U(4) groups. This similarity justifies our definition of a U(4) Layer groups for *each* of the four generations yielding a new 'composite' Layer group $U(4)^4$. *Each generation in the four layers has a U(4) transformation symmetry group mixing the fermions in the generation across all four layers.*

2.1.2 Layer Group Particle Layers

It is important to note that the Layer numbers are independent of the baryon and lepton particle numbers that form the basis of the Generation group, and so the physics embodied in the Generation group is not the same as the physics of the Layer groups defined here.

Layer numbers are conserved under strong and electromagnetic interactions but broken by the Electromagnetic and Weak interactions as well as their Dark counterparts.

Since the coefficients in Layer group transformations can be local functions, the new Layer groups are implemented with Yang-Mills fields.

Just as we extended the reach of the Generation group from one generation to four generations, we can extend each of the Layer groups representations to 4 similarly by assuming that there are four layers of fermions for each generation. We add a U(4) layer index to each fermion field for each of the four Layer groups. Fig. 2.2 shows the four U(4) Layer groups rotations of each of the four generation's fermions.

Further the gauge fields for SU(3)⊗SU(2)⊗U(1)⊗SU(2)⊗U(1)⊗U(4) must now be different for each layer.Thus interactions of these types between fermions in different layers is prevented.

To implement this new Layer symmetry we expand the Extended Standard Model lagrangian the following steps to obtain the SuperStandard Model:

1, All covariant derivatives must expand to four U(4) Layer gauge fields terms. The new terms are for four layers of SU(3)⊗SU(2)⊗U(1)⊗SU(2)⊗U(1)⊗U(4) fields. Each gauge field then has an additionl index specifying its layer. The rationale for the choice of these sets of groups to be duplicated is that they, and only they, all play a necessary role in determining the structure of the Periodic Table of Fermions.

2. The new Layer groups are the same as in the Extended Standard Model of volume I. Each fermion particle field has an index labeling the layer of the particle making four layers of four generations of fermions.

3. Each layer should have its own set of Higgs particles (modulo mixing) contributing to fermion masses. A layer index number must be added to each Higgs field. One expects that the masses of fermions should be substantially larger for the three 'upper' layers beyond our layer.[26] Otherwise we would have found particles from these upper layers.

The above modifications to the Extended Standard Model lagrangian result in the Unified SuperStandard Model. The form of the "periodic tables" of fermion and vector bosons that results appears in Figs. 2.3 and 2.4. The 'splitting' of a single fermion generation into four grneratios and then into four layers is shown in Fig. 2.5.

2.2 Layer Group Features

The new Layer groups have a number of features that follow from the Layer group discussions detailed in volume I. These features can easily be modified to the new form of the Layer groups. They include Higgs contributions to gauge field masses, the four layer CKM-like matrix, and new Layer group gauge field interactions.

2.3 New Labeling of Fermion Periodic Table Particles

Fig. 2/6 has the Periodic Table of Fermions with rows and columns labeled with quadruplets of numbers. In the case of normal quark species, which each consist of a color triplet, a fifth integer would be needed to identify each color quark within a triplet: (T, S, L, G, C) where C = 1, 2, 3 identifies the color ("red, white or blue"). For example, the second generation, fourth layer, normal up-type red quark is (1, 3, 4, 2, 1) where we treat 'red' as having the value 1.

[26] The interplayed mixing of the particles in each generation between different layers may partly explain the vast increases seen in fermion masses as one goes from generation to generation in each species. The Higgs particles in different layers are different and a possible source of the growth of fermion masses.

THE FERMION PERIODIC TABLE

Figure 2.3. Dark parts of the periodic table are gray. Light parts are the known fermions with an additional, as yet not found, 4th generation shown.

THE VECTOR BOSON PERIODIC TABLE

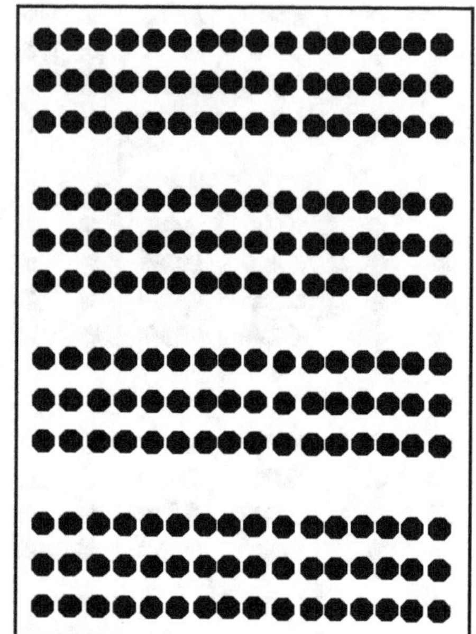

Figure 2.4. The known vector bosons are in the lowest row. The Layer groups are distributed by layer symbolically although they each straddle all four layers.

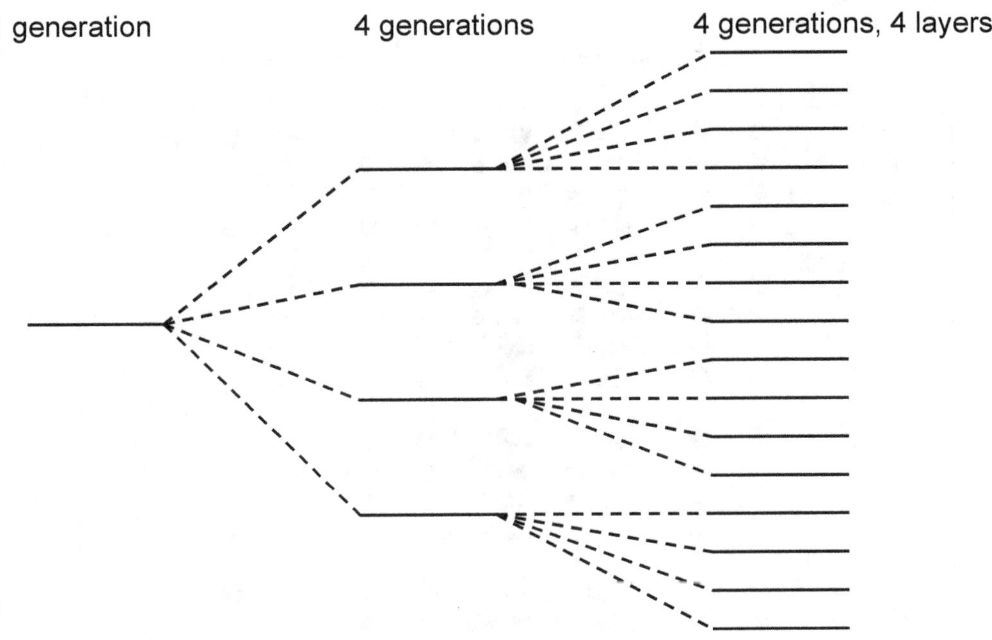

Figure 2.5. The 'splitting of a single generation fermion into four generations and then into four layers.

However, only the color SU(3) interaction distinguishes between colors in interactions. Thus we can use quadruplets for all practical purposes – for electromagnetic, Weak, Generation group, and Layer interactions of fermions.

The quadruplet numbering patern is as follows. Species are numbered from 1 through 4 – separately for normal and Dark species: charged lepton, neutral lepton, three up-type quarks, three down-type quarks, Dark charged lepton, Dark neutral lepton, one Dark up-type quark species, and one Dark down-type quark species. Layers are numbered from 1 through 4 with our layer being layer 1. Generations are numbered from 1 through 4 from lightest to the heaviest.[27] (e, v_e, u and d constitute the known part of generation 1 of our layer.) We will call the quadruplet (or quintet) of a fermion its *ID number*. See Fig. 2.6.

[27] The ordering by mass may not hold in the currently Dark part of the fermion spectrum.

Figure 2.6. The Periodic Table of Fermions with each fermion identified by a quadruplet of integers: (T, S, L, G) where T identifies Normal or Dark, S identifies the species, G identifies the generation, and L identifies the layer. For example the electron has the triplet (1, 1, 1, 1), and the second generation, fourth layer, Dark up-type quark is (2, 3, 4, 2).

Since the Periodic table of Fermions is 2-dimensional – like the Chemical Periodic Table of Elements one might ask: Why not use two numbers to identify fermions in a manner similar to the Chemistry table? The answer is related to the group structure of the Extended Standatd Model interactions. Except for T (the type of matter), the other three digits in a quadruplet are each related to a group. Thus G specifies the generation and thus the position of a fermion in the 4 of the U(4) Generation group. The integer L specifies the layer and thus the position of the fermion in the 4 of the volume I form of the U(4) Layer group.

The integer S specifies the position of a fermion amongst the species. In volume I we saw that the four species of each type of matter follow from a U(4) group, that we called the General Relativistic Reality Group which was derived from a consideration of the structure of Complex General Coordinate transformations. The U(4) gauge fields are denoted $A_S{}^\mu(x)$ in discussions below.

Thus the fermion quadruplet labeling is physically motivated by the group structure of the SuperStandard Model: $[SU(3) \otimes SU(2) \otimes U(1) \otimes SU(2) \otimes U(1) \otimes U(4) \otimes U(4)]^4 \otimes U(192)$, which we will explain in detail later.

2.4 Equipartition Principle for Fermions and Gauge Fields

We now[28] will suggest a rationale for the dominant abundance of Dark mass-energy:

2.4.1 Equipartition Principle for Particle Degrees of Freedom

In a closed system at equilibrium the thermal energy of a system is equally partitioned (distributed) among its degrees of freedom. This Equipartition Principle is well known. The application of this principle to the beginning of the universe *when all particles were massless* and all symmetries were unbroken suggests that the distribution of mass-energy should be the same for all degrees of freedom at that time. Thus there should be approximately equal numbers and energies of 192 fermions and 192 vector bosons with the same fraction of the total thermal energy.

We now estimate the relative proportion of Normal and Dark matter in the universe at its beginning based on this Equipartition Principle.

2.4.2 Proportion of Dark Mass-Energy in the Universe

First we note that 8 of the 12 fermion species (counting quark of each color as a separate species) in layer 1 – the layer with which we are familiar are Normal matter fermions. (Our discussion is based on our SuperStandard Model.) Four of the 12 first layer species are Dark.

The other three fermion layers are all Dark from our point of view since they have not been detected. Thus we find that 40 of the 48 species are Dark yielding a *percentage of Dark Matter equal to 40/48 = 83.33%*. (The same counting could have been done by counting individual fermions with the same results.)

Recent studies of the proportion of Dark Matter in the universe have yielded two estimates: 84.5% by Aghanim et al in Astronomy and Astrophysics 1303;5062 and 81.5% from a NASA fit to various models.

Thus our estimate based on our fermion Equipartition Principle is midway between these experimental estimates.

Two possibilities emerge with respect to the present proportion of Dark Matter:

1. The percentage has not changed from the Beginning and the approximate estimates are slightly off. The lack of change could be due to the extremely small decay rates of the fermions in the higher layers.

2. The percentage of matter in the upper layers has decreased due to decay and so the

[28] Some of the material in this section appeared in Blaha (2016a).

current proportion may be somewhat below 83.33%.

We know of 12 of the 192 vector bosons in the SuperStandard Model and 24 fermions. Thus we find a total of 348 out of 384 particles are Dark yielding a Dark mass-energy of 91% of the universes mass-energy at the beginning of the universe.

The sum of Dark energy in the universe currently has been estimated to be 68% of the total energy and the energy of Dark Matter is estimated to be 26.8%. The total is 95% - a value that compares favorably with our above approximate estimate of 91%. The dfference in these values can be attributed to various factors. One distinct possibility is the decay of Dark mass-energy from higher layers to the known layer in the 13.8 billion years since the Big Bang.

2.5 The New 'Interaction Rotation' Interaction A_Θ

In volume I we defined an 'Interaction Rotation' group, the SU(3)⊗U(64) Ω-Symmetry group, with gauge fields denoted $A_\Omega{}^\mu(x)$.

We now define a new 'Interaction Rotation' group, the Θ-Symmetry group G_Θ. It rotates the 192 interaction gauge field components of

$$\Theta = \prod_{k=1}^{4} SU(3)_k \otimes SU(2)_k \otimes U(1)_k \otimes SU(2)_k \otimes U(1)_k \otimes U(4)_k \otimes U(4)_k \tag{2.2}$$

where k labels the layer for all factors except the last U(4) factor where it enumerates the four Layer groups defined above[29]. We can abbreviate Θ as

$$\Theta = [SU(3) \otimes SU(2) \otimes U(1) \otimes SU(2) \otimes U(1) \otimes U(4) \otimes U(4)]^4 \tag{2.3}$$

Each layer has its own set of Standard Model-like gauge fields plus there are four Layer group gauge fields that operate between layers. Thus there is no direct leakage between layers except for the 'Interaction Rotation' group which rotates all the components of all of the fields in Θ. Since G_Θ performs complex transformations of the 192 components of the gauge fields of Θ, we see that $G_\Theta = U(192)$.

The 192 gauge fields that are rotated, using vector notation, are

$$\mathbf{A}_I{}^\mu = \Sigma_a{}^a\mathbf{A}_I{}^\mu(x) \cdot {}^a\mathbf{T}_I \tag{2.4}$$

where ${}^a\mathbf{T}_I$ is given below in eq. 2.5 and

$${}^a\mathbf{A}_I{}^\mu = (g^a{}_1{}^a\mathbf{A}_{SU(3)}{}^\mu(x_C),\ g^a{}_2{}^a\mathbf{W}^\mu(x)\ ,\ g^a{}_3{}^a\mathbf{A}_E{}^\mu(x),\ g^a{}_4{}^a\mathbf{W}_D{}^\mu(x),\ g^a{}_5{}^a\mathbf{A}_{DE}{}^\mu(x),\ g^a{}_6{}^a\mathbf{U}^\mu(x)\ ,\ g^a{}_7{}^a\mathbf{V}^\mu(x)) \tag{2.4a}$$

[29] These groups appear in the Θ-symmetry group because they are all directly connected to defining the form of the Periodic Table of Fermions. The other groups do not specify aspects of the form.

for a = 1,2, 3, 4 which specifies the layer[30]. Each vector $^a\mathbf{A}_I{}^\mu$ is a vector of the gauge fields of the respective group in layer a. For simplicity we label the respective layer dependent coupling constants as $g^a{}_1$, $g^a{}_2$, … , $g^a{}_7$. In eq. 2.4 the subscript 'D' labels Dark matter interactions, 'W' labels Weak fields, 'E' labels Electromagnetic fields, $U^\mu(x)$ labels U(4) Generation group fields, and $V^\mu(x)$ labels U(4) Layer group fields.

　　Similarly we define a 7-vector of 48 matrix generators for each a:

$$^a\mathbf{T}_I = (^a\mathbf{T}_{SU(3)},\ ^a\boldsymbol{\tau}_{SU(2)},\ ^a\mathbf{I}_{U(1)},\ ^a\boldsymbol{\tau}_{DSU(2)},\ ^a\mathbf{I}_{DU(1)},\ ^a\mathbf{G}_{U(4)}),\ ^a\mathbf{G}_{LU(4)}) \qquad (2.5)$$

We put a layer index on these matrices: since, although they have the same form in all layers, the respective matrices operate in 'different spaces' for each layer since the gauge fields, and the fermion fields they operate on, differ from layer to layer.

　　The plenitude of interactions that we have identified and summarized in chapter 25 of volume I leads us to consider the possibility of unification based partly on the 'rotation of interactions' and more fully, on a unification based on the Riemann-Christoffel tensor.[31]

　　When we use the Pseudoquantum (two fields per gauge particle) formalism[32] the part of these gauge fields interactions within a covariant derivative has the form

$$^a\mathbf{A}_I{}^{1\mu}(x)\cdot{}^a\mathbf{T}_I + {}^a\mathbf{A}_I{}^{2\mu}(x)\cdot{}^a\mathbf{T}_I = {}^a\mathbf{A}_I{}^{1\mu}{}_k(x){}^a\mathbf{T}_{Ik} + {}^a\mathbf{A}_I{}^{2\mu}{}_k(x){}^a\mathbf{T}_{Ik} \qquad (2.6)$$

summed over a and k.

2.5.1 Θ-Symmetry, 'Interaction Rotations' for Fermions

　　The Θ-Symmetry gauge fields appear in Dirac-like equation covariant derivatives.We define a column vector ψ containing the 4-spinors of all 192 normal and Dark fundamental fermions in our theory based on four generations in four layers as detailed earlier and in Blaha (2016a), (2016b) and (2016c). Then the flat space-time Dirac equation[33] has the form:

$$\gamma_\mu D^\mu\psi = \gamma_\mu\{\partial^\mu + i\,[g_\Theta A_\Theta{}^{1\mu}(x) + g_\Theta A_\Theta{}^{2\mu}(x) + \Sigma_a(^a\mathbf{A}_I{}^{1\mu}(x)\cdot{}^a\mathbf{T}_I + {}^a\mathbf{A}_I{}^{2\mu}(x)\cdot{}^a\mathbf{T}_I)]\}\psi = 0 \qquad (2.7)$$

with the Θ-Symmetry gauge field terms $A_\Theta{}^{1\mu}(x)$ and $A_\Theta{}^{2\mu}(x)$ inserted. Each of these fields consists of a gauge field multiplied by an U(192) matrix generator in the <u>192</u> representation since there are 192 fermion fields.

　　We now define

[30] In the case of the Layer groups' gauge fields $U^\mu(x)$ the index 'a' merely numbers the four Layer groups since the groups straddle all four layers.

[31] Much of this chapter first appeared in Blaha (2017a).

[32] See volume I.

[33] Similar considerations apply to other Dirac-like equations described in volume I.

$$\mathbf{A}_I{}^{i\mu} = \Sigma_a {}^a\mathbf{A}_I{}^{i\mu}(x) \cdot {}^a\mathbf{T}_I \tag{2.8}$$

for $i = 1, 2$ since we will be performing Θ transformations that will mix all the field components within $\mathbf{A}_I{}^{i\mu}$. Then the Dirac equation becomes

$$\gamma_\mu D^\mu \psi = \gamma_\mu \{\partial^\mu + i\,[g_\Theta A_\Theta{}^{1\mu}(x) + g_\Theta A_\Theta{}^{2\mu}(x) + \mathbf{A}_I{}^{1\mu}(x) + \mathbf{A}_I{}^{2\mu}(x)]\}\psi = 0 \tag{2.9}$$

The $\mathbf{A}_\Theta{}^{1\mu}(x)$ and $\mathbf{A}_\Theta{}^{2\mu}(x)$ gauge fields each separately have a 192×192 matrix representation:

$$\mathbf{A}_\Theta{}^{i\mu}(x) = \Sigma_n \mathbf{A}_\Theta{}^{i\mu}{}_n(x) T_{U(192)n} \tag{2.10}$$

for $i = 1., 2$ where n sums over the 192^2 generators of U(192).

The fields in $A_I{}^{1\mu}(x)$, and $A_I{}^{2\mu}(x)$ separately transform under a Θ-transformation. Under a Θ-transformation we find the Dirac equation eq. 2.9 transforms to

$$\gamma_\mu\{\partial^\mu + i\,[g_\Theta A'_\Theta{}^{1\mu}(x) + g_\Theta A'_\Theta{}^{2\mu}(x) + \mathbf{A}'_I{}^{1\mu}(x) + \mathbf{A}'_I{}^{2\mu}(x)]\}\psi' = 0 \tag{2.11}$$

where

$$A'_\Theta{}^{1\mu}(x) = C_\Theta(x)A_\Theta{}^{1\mu}(x)C_\Theta{}^{-1}(x) - i\,C_\Theta(x)\partial^\mu C_\Theta{}^{-1}(x)/g_\Theta \tag{2.12}$$
$$A'_\Theta{}^{2\mu}(x) = C_\Theta(x)A_\Theta{}^{2\mu}(x)C_\Theta{}^{-1}(x)$$
$$\mathbf{A}'_I{}^{1\mu}(x) = C_\Theta(x)\mathbf{A}_I{}^{1\mu}(x)\,C_\Theta{}^{-1}(x)$$
$$\mathbf{A}'_I{}^{2\mu}(x) = C_\Theta(x)\mathbf{A}_I{}^{2\mu}(x)\,C_\Theta{}^{-1}(x)$$
$$\psi' = C_\Theta(x)\psi$$

The effect of the Θ-transformation is to rotate the gauge fields components. It is accompanied by a rotation of the generator matrices components. Together they define an equivalent formulation of the original Dirac equation and thus the fermion sector. The next section provides an explicit simple example of Θ-transformations – an ElectroWeak theory.

Note that an Θ-transformation causes a change of gauge in the $A_\Theta{}^{1\mu}(x)$ field. The other gauge fields, $A_\Theta{}^{2\mu}(x)$, $\mathbf{A}_I{}^{1\mu}(x)$ and $\mathbf{A}_I{}^{2\mu}(x)$, are 'rotated' but do not undergo a change of gauge. (Each of the 28 gauge fields within $\mathbf{A}_I{}^{1\mu}(x)$ and $\mathbf{A}_I{}^{2\mu}(x)$ do undergo their own particular changes of gauge for their transformation groups.)

2.5.2 Broken Θ-Symmetry

Θ-symmetry is broken by the complexon nature of color SU(3) gauge fields and quarks. The color SU(3) gauge fields $A_{SU(3)}{}^{j\mu}(x)$ for $j = 1,2$ appearing in $\mathbf{A}_I{}^{i\mu}(x)$ are complexon fields in our formulation of SuperStandard Model. Thus they are functions of complex spatial coordinates

$$x_c = (t,\, \mathbf{x}_r - i\mathbf{x}_i) \tag{2.13}$$

Therefore color SU(3) does not admit of Θ-transformations. Consequently Θ-Symmetry is broken to $[SU(3)_{Color}]^4 \otimes U(160)$. This symmetry breaking reduces the number of Θ-Symmetry generators to 160^2.

The symmetry breaking mechanism may be expected to lead to large A_Θ gauge field masses. Together with the likely smallness of the g_Θ coupling constant, this leads us to expect that the A_Θ interaction, which occurs between all 192 fermions, will not be detectable.

2.6 Example: ElectroWeak-like Theory

ElectroWeak model is an example of a global Θ-transformation which does not include the full gamut of features outlined above. Focussing on the Weak and Electromagnetic interactions we consider the covariant derivative

$$\{\partial^\mu + i\,[g\mathbf{W}^\mu\cdot\boldsymbol{\tau} + g'W_0{}^\mu\tau_0]\}\psi = 0 \tag{2.14}$$

Rotating $W_3{}^\mu$ and $W_0{}^\mu$ with $C^{-1}{}_\Theta$

$$\begin{bmatrix} gW_3{}^\mu \\[2mm] g'W_0{}^\mu \end{bmatrix} \rightarrow \begin{bmatrix} g\cos\theta\, Z^\mu + g\sin\theta\, A^\mu \\[2mm] -g'\sin\theta\, Z^\mu + g'\cos\theta\, A^\mu \end{bmatrix}$$

and rotating the generator matrices by C_Θ

$$\begin{bmatrix} \tau_3 \\[2mm] \tau_0 \end{bmatrix} \rightarrow \begin{bmatrix} \cos\theta\,\tau_3 - \sin\theta\,\tau_0 \\[2mm] \sin\theta\,\tau_3 + \cos\theta\,\tau_0 \end{bmatrix} \tag{2.15}$$

and making an appropriate choice of θ_W yields the electromagnetic field A and Z we find

$$g'W_0{}^\mu \tau_0 + gW_3{}^\mu\tau_3 \rightarrow Z^\mu[(g\cos^2\theta - g'\sin^2\theta)\tau_3 - \tfrac{1}{2}\sin(2\theta)(g+g')\tau_0 + A^\mu[\tfrac{1}{2}\sin(2\theta)(g+g')\tau_3 + (g'\cos^2\theta - g\sin^2\theta)\tau_0] \tag{2.16}$$

If we choose the coefficient of A^μ be e, then

$$e(\tau_3 + \tau_0) = [\tfrac{1}{2}\sin(2\theta)(g+g')\tau_3 + (g'\cos^2\theta - g\sin^2\theta)\tau_0] \tag{2.17}$$

and thus

$$\tfrac{1}{2}\sin(2\theta)(g + g') = (g'\cos^2\theta - g\sin^2\theta) = e \qquad (2.18)$$

Consequently

$$g' = -(\tfrac{1}{2}\sin(2\theta) - \sin^2\theta)/(\tfrac{1}{2}\sin(2\theta) - \cos^2\theta) \qquad (2.19)$$

If we further choose $g = g'$ (since equal coupling constants is an often stated goal of theorists) for simplicity we find the angle $\theta = \pi/8$ or $22.5°$ while the usual Weinberg angle θ_W is about $30°$. Thus we find a close similarity to standard ElectroWeak theory.[34]

We conclude that ElectroWeak theory can be viewed as an example of a Θ-transformation.[35]

2.7 Θ-Symmetry and Higgs Fields

The Higgs boson fields η (or our Pseudoquantum alternative) also participate in Θ-symmetry rotations. Consider a composite boson field constructed from the concatenation of all Higgs fields. The term generating the gauge field masses has the form

$$(D^\mu\eta)^\dagger D^\mu\eta \qquad (2.20)$$

Upon rotating gauge fields, mass terms will also correspondingly 'rotate' to maintain the covariance of the overall theory.

The A_Θ gauge fields will also all acquire masses since there is no evidence of long range A_Θ fields experimentally.

2.8 Path Integral Formulation, and the Faddeev-Popov Method

The path integral formulation of our theory, with the complete set of eleven integrations described in volume I, is fairly straightforward with one exception. Since we use complex valued coordinates for the color SU(3) gauge theory a somewhat different approach must be followed for it. We detail this approach in Blaha (2015a).

The Faddeev-Popov path integral formulation is the same as that of the $A_\Omega^{1\mu}(x)$ path integral formulation presented in volume I with only minor notational differences.

2.9 SuperStandard Covariant Derivative

The covariant derivative[36] which appears in fermion and gravitation equations uses[37]

[34] We can, of course, make the angles equal by adjusting the values of g ang g'.

[35] A rotation of fermions was not required in this case because the interactions are not charge changing. In the general case a rotation of interactions also requires a rotation of the 192 fermions.

[36] This section has equations obtained from volume I. The equation numbering is that of volume I.

[37] From eqs. 2.4a and 2.8.

$$^a\mathbf{A}_I{}^{i\mu} = (g^a{}_1{}^a\mathbf{A}_{SU(3)}{}^{i\mu}(x_C),\ g^a{}_2{}^a\mathbf{W}^{i\mu}(x),\ g^a{}_3{}^a\mathbf{A}_E{}^{i\mu}(x),\ g^a{}_4{}^a\mathbf{W}_D{}^{i\mu}(x),\ g^a{}_5{}^a\mathbf{A}_{DE}{}^{i\mu}(x),\ g^a{}_6{}^a\mathbf{U}^{i\mu}(x),\ g^a{}_7{}^a\mathbf{V}^{i\mu}(x))$$

$$(2.21)$$

where each element is a vector of the gauge fields[38] of the respective group for $i = 1, 2$ and

$$\mathbf{A}_I{}^{i\mu} = \Sigma_a\ {}^a\mathbf{A}_I{}^{i\mu}(x)\cdot{}^a\mathbf{T}_I + g_8\mathbf{A}_S{}^\mu(x) \qquad (2.22)$$

where a labels the layer.

For simplicity we label the respective coupling constants as $g_1, g_2, \ldots, g_8$. In eq. 2.21 above the subscript 'D' labels Dark matter interactions, 'W' labels Weak fields, 'E' labels Electromagnetic fields, $V^\mu(x)$ labels U(4) Generation group fields, and 'V' labels U(4) Layer group fields. A_S labels the U(4) General Relativistic transformations Reality field.

The interactions' symmetry is $[SU(3)\otimes SU(2)\otimes U(1)\otimes SU(2)\otimes U(1)\otimes U(4)\otimes U(4)]^4$ respectively. The number of gauge fields totals to 192. There is also the U(192) 'rotations of interactions' set of gauge fields as well as gravitation terms.

Using the above definitions the covariant derivative of a 4-vector is

$$
\begin{aligned}
D_\nu V_\mu &= (\partial_\nu + iF_\nu)V_\mu - H^\sigma{}_{\nu\mu}V_\sigma \\
&= [g^\sigma{}_\mu \partial_\nu + ig^\sigma{}_\mu F_\nu - H^\sigma{}_{\nu\mu}]V_\sigma \\
&= [g^\sigma{}_\mu \partial_\nu + iD^\sigma{}_{\mu\nu}]V_\sigma
\end{aligned}
\qquad (2.23)
$$

where[39]

$$F^\mu = g_\Theta A_\Theta{}^{1\mu}(x) + g_\Theta A_\Theta{}^{2\mu}(x) + \mathbf{A}_I{}^{1\mu}(x) + \mathbf{A}_I{}^{2\mu}(x) + B^{1\mu} + B^{2\mu} \qquad (2.24)$$

using eq. 2.22, and

$$H^\sigma{}_{\nu\mu} = \Gamma_{GR}{}^\sigma{}_{\nu\mu} + \Gamma_{GR}{}^{2\sigma}{}_{\nu\mu} \qquad (2.25)$$

$$D^\sigma{}_{\mu\nu} = g^\sigma{}_\mu F_\nu + iH^\sigma{}_{\nu\mu} \qquad (2.26)$$

where we have abstracted the complex part of the complex affine connection into the U(4) gauge field $A_S{}^\mu$. Eq. 2.25 is the real-valued part of the complex affine connection.

Commutators of the vector fields in F_μ are implicit when the covariant derivative is applied to vectors and tensors such as V_σ.

2.10 SuperStandard Curvature Tensor

The curvature tensor (applied to a 4-vector) defined in volume I is

$$
\begin{aligned}
R'^\beta{}_{\sigma\nu\mu}V_\beta = \ &g^\alpha{}_\mu(\partial_\nu + iF_\nu)g^\beta{}_\sigma(\partial_\alpha + iF_\alpha)V_\beta - H^\alpha{}_{\mu\nu}g^\beta{}_\sigma(\partial_\alpha + iF_\alpha)V_\beta + \\
&+ H^\alpha{}_{\mu\nu}H^\beta{}_{\sigma\alpha}V_\beta - g^\alpha{}_\mu(\partial_\nu + iF_\nu)H^\beta{}_{\sigma\alpha}V_\beta - H^\gamma{}_{\nu\sigma}\{g^\alpha{}_\gamma(\partial_\mu + iF_\mu)V_\alpha - H^\alpha{}_{\gamma\mu}V_\alpha\} - \\
&- \{\mu \leftrightarrow \nu\}
\end{aligned}
$$

[38] We use Pseudoquantum fields with two sets of fields per vector boson.
[39] We will omit the insertion of coupling constants of $B^{1\mu}$ and $B^{2\mu}$ in the interests of simplifying expressions.

$$= ig^\beta{}_\sigma(\partial_\nu F_\mu - \partial_\mu F_\nu - i[F_\nu, F_\mu])V_\beta + (\partial_\mu H^\beta{}_{\sigma\nu} - \partial_\nu H^\beta{}_{\sigma\mu} + H^\gamma{}_{\nu\sigma}H^\beta{}_{\gamma\mu} - H^\gamma{}_{\mu\sigma}H^\beta{}_{\gamma\nu})V_\beta$$

$$= ig^\beta{}_\sigma(F_E{}^1{}_{\nu\mu} + F_E{}^2{}_{\nu\mu} + F_W{}^1{}_{\nu\mu} + F_W{}^2{}_{\nu\mu} + F_{DE}{}^1{}_{\nu\mu} + F_{DE}{}^2{}_{\nu\mu} + F_{DW}{}^1{}_{\nu\mu} + F_{DW}{}^2{}_{\nu\mu} +$$
$$+ F_{SU(3)}{}^1{}_{\nu\mu} \; F_{SU(3)}{}^2{}_{\nu\mu} + F_U{}^1{}_{\nu\mu} + F_U{}^2{}_{\nu\mu} + F_V{}^1{}_{\nu\mu} + F_V{}^2{}_{\nu\mu} + F_\Theta{}^1{}_{\nu\mu} + F_\Theta{}^2{}_{\nu\mu})V_\beta +$$
$$+ (ig^\beta{}_\sigma B^1{}_{\nu\mu} + ig^\beta{}_\sigma B^2{}_{\nu\mu} + \partial_\mu H^\beta{}_{\sigma\nu} - \partial_\nu H^\beta{}_{\sigma\mu} + H^\gamma{}_{\nu\sigma}H^\beta{}_{\gamma\mu} - H^\gamma{}_{\mu\sigma}H^\beta{}_{\gamma\nu})V_\beta$$

$$= R'_E{}^\beta{}_{\sigma\nu\mu}V_\beta + R'_{SU(2)}{}^\beta{}_{\sigma\nu\mu}V_\beta + R'_{DE}{}^\beta{}_{\sigma\nu\mu}V_\beta + R'_{DSU(2)}{}^\beta{}_{\sigma\nu\mu}V_\beta + R'_{SU(3)}{}^\beta{}_{\sigma\nu\mu}V_\beta + R'_U{}^\beta{}_{\sigma\nu\mu}V_\beta +$$
$$+ R'_V{}^\beta{}_{\sigma\nu\mu}V_\beta + R'_S{}^\beta{}_{\sigma\nu}V_\beta + R'_\Theta{}^\beta{}_{\sigma\nu}V_\beta + R'_B{}^\beta{}_{\sigma\nu}V_\beta + R'_G{}^\beta{}_{\sigma\nu\mu}V_\beta \qquad (2.27)$$

where the $R'^\beta{}_{\sigma\nu\mu}$ below for each of the interactions listed in eq. 2.21 is a sum of $R'^\beta{}_{\sigma\nu\mu}$ terms over the layers

$$R'_{SU(3)}{}^\beta{}_{\sigma\nu\mu} = ig^\beta{}_\sigma(F_{SU(3)}{}^1{}_{\nu\mu} + F_{SU(3)}{}^2{}_{\nu\mu}) \qquad (2.28)$$
$$R'_{SU(2)}{}^\beta{}_{\sigma\nu\mu} = ig^\beta{}_\sigma(F_W{}^1{}_{\nu\mu} + F_W{}^2{}_{\nu\mu})$$
$$R'_E{}^\beta{}_{\sigma\nu\mu} = ig^\beta{}_\sigma(F_E{}^1{}_{\nu\mu} + F_E{}^2{}_{\nu\mu})$$
$$R'_U{}^\beta{}_{\sigma\nu\mu} = ig^\beta{}_\sigma(F_U{}^1{}_{\nu\mu} + F_U{}^2{}_{\nu\mu})$$
$$R'_V{}^\beta{}_{\sigma\nu\mu} = ig^\beta{}_\sigma(F_V{}^1{}_{\nu\mu} + F_V{}^2{}_{\nu\mu})$$
$$R'_{DSU(2)}{}^\beta{}_{\sigma\nu\mu} = ig^\beta{}_\sigma(F_{DW}{}^1{}_{\nu\mu} + F_{DW}{}^2{}_{\nu\mu})$$
$$R'_{DE}{}^\beta{}_{\sigma\nu\mu} = ig^\beta{}_\sigma(F_{DE}{}^1{}_{\nu\mu} + F_{DE}{}^2{}_{\nu\mu})$$
$$R'_S{}^\beta{}_{\sigma\nu\mu} = ig^\beta{}_\sigma(F_S{}^1{}_{\nu\mu} + F_S{}^2{}_{\nu\mu})$$
$$R'_\Theta{}^\beta{}_{\sigma\nu\mu} = ig^\beta{}_\sigma(F_\Theta{}^1{}_{\nu\mu} + F_\Theta{}^2{}_{\nu\mu})$$
$$R'_B{}^\beta{}_{\sigma\nu\mu} = ig^\beta{}_\sigma(F_B{}^1{}_{\nu\mu} + F_B{}^2{}_{\nu\mu})$$

and

$$R'_G{}^\beta{}_{\sigma\nu\mu} = \partial_\mu H^{1\beta}{}_{\sigma\nu} - \partial_\nu H^{1\beta}{}_{\sigma\mu} + H^{1\gamma}{}_{\nu\sigma}H^{1\beta}{}_{\gamma\mu} - H^{1\gamma}{}_{\mu\sigma}H^{1\beta}{}_{\gamma\nu} + \partial_\mu H^{2\beta}{}_{\sigma\nu} - \partial_\nu H^{2\beta}{}_{\sigma\mu} +$$
$$+ H^{2\gamma}{}_{\nu\sigma}H^{2\beta}{}_{\gamma\mu} - H^{2\gamma}{}_{\mu\sigma}H^{2\beta}{}_{\gamma\nu} + H^{1\gamma}{}_{\nu\sigma}H^{2\beta}{}_{\gamma\mu} - H^{1\gamma}{}_{\mu\sigma}H^{2\beta}{}_{\gamma\nu} + H^{2\gamma}{}_{\nu\sigma}H^{1\beta}{}_{\gamma\mu} - \Gamma^{2\gamma}{}_{\mu\sigma}\Gamma^\beta{}_{\gamma\nu} \qquad (2.29)$$
$$= R^{1\beta}{}_{\sigma\nu\mu} + R^{2\beta}{}_{\sigma\nu\mu}$$

with

$$H^\beta{}_{\sigma\nu\mu} = \partial_\mu H^\beta{}_{\sigma\nu} - \partial_\nu H^\beta{}_{\sigma\mu} + H^\gamma{}_{\nu\sigma}H^\beta{}_{\gamma\mu} - H^\gamma{}_{\mu\sigma}H^\beta{}_{\gamma\nu} \qquad (2.30)$$
$$R^{1\beta}{}_{\sigma\nu\mu} = \partial_\mu H^{1\beta}{}_{\sigma\nu} - \partial_\nu H^{1\beta}{}_{\sigma\mu} + H^{1\gamma}{}_{\nu\sigma}H^{1\beta}{}_{\gamma\mu} - H^{1\gamma}{}_{\mu\sigma}H^{1\beta}{}_{\gamma\nu} \qquad (2.31)$$
$$R^{2\beta}{}_{\sigma\nu\mu p} = \partial_\mu H^{2\beta}{}_{\sigma\nu} - \partial_\nu H^{2\beta}{}_{\sigma\mu} + H^{2\gamma}{}_{\nu\sigma}H^{2\beta}{}_{\gamma\mu} - H^{2\gamma}{}_{\mu\sigma}H^{2\beta}{}_{\gamma\nu} +$$
$$+ H^{1\gamma}{}_{\nu\sigma}H^{2\beta}{}_{\gamma\mu} - H^{1\gamma}{}_{\mu\sigma}H^{2\beta}{}_{\gamma\nu} + H^{2\gamma}{}_{\nu\sigma}H^{1\beta}{}_{\gamma\mu} - H^{2\gamma}{}_{\mu\sigma}H^{1\beta}{}_{\gamma\nu} \qquad (2.32)$$

and

$$H^{1\sigma}{}_{\nu\mu} = \Gamma_{GR}{}^\sigma{}_{\nu\mu}$$
$$H^{2\sigma}{}_{\nu\mu} = \Gamma_{GR}{}^{2\sigma}{}_{\nu\mu}$$

and where

$$F_{SU(3)}{}^1{}_{\varkappa\mu} = \partial A_{SU(3)}{}^1{}_{\mu}/\partial x^{\varkappa} - \partial A_{SU(3)}{}^1{}_{\varkappa}/\partial x^{\mu} + ig_1[A_{SU(3)}{}^1{}_{\varkappa}, A_{U(3)}{}^1{}_{\mu}] \qquad (2.33)$$

$$F_{W}{}^1{}_{\varkappa\mu} = \partial W^1{}_{\mu}/\partial x^{\varkappa} - \partial W^1{}_{\varkappa}/\partial x^{\mu} + ig_2[W^1{}_{\varkappa}, W^1{}_{\mu}]$$

$$F_{E}{}^1{}_{\varkappa\mu} = \partial A_{E}{}^1{}_{\mu}/\partial x^{\varkappa} - \partial A_{E}{}^1{}_{\varkappa}/\partial x^{\mu}$$

$$F_{DW}{}^1{}_{\varkappa\mu} = \partial W_{D}{}^1{}_{\mu}/\partial x^{\varkappa} - \partial W_{D}{}^1{}_{\varkappa}/\partial x^{\mu} + ig_4[W_{D}{}^1{}_{\varkappa}, W_{D}{}^1{}_{\mu}]$$

$$F_{DE}{}^1{}_{\varkappa\mu} = \partial A_{DE}{}^1{}_{\mu}/\partial x^{\varkappa} - \partial A_{DE}{}^1{}_{\varkappa}/\partial x^{\mu}$$

$$F_{U}{}^1{}_{\varkappa\mu} = \partial U^1{}_{\mu}/\partial x^{\varkappa} - \partial U^1{}_{\varkappa}/\partial x^{\mu} + ig_6[U^1{}_{\varkappa}, U^1{}_{\mu}]$$

$$F_{V}{}^1{}_{\varkappa\mu} = \partial V^1{}_{\mu}/\partial x^{\varkappa} - \partial V^1{}_{\varkappa}/\partial x^{\mu} + ig_7[V^1{}_{\varkappa}, V^1{}_{\mu}]$$

$$F_{S}{}^1{}_{\varkappa\mu} = \partial A_{S}{}^1{}_{\mu}/\partial x^{\varkappa} - \partial A_{S}{}^1{}_{\varkappa}/\partial x^{\mu} + ig_8[A_{S}{}^1{}_{\varkappa}, A_{S}{}^1{}_{\mu}]$$

$$F_{\Theta}{}^1{}_{\varkappa\mu} = \partial A_{\Theta}{}^1{}_{\mu}/\partial x^{\varkappa} - \partial A_{\Theta}{}^1{}_{\varkappa}/\partial x^{\mu} + ig_9[A_{\Theta}{}^1{}_{\varkappa}, A_{\Theta}{}^1{}_{\mu}]$$

$$F_{B}{}^1{}_{\varkappa\mu} = \partial B^1{}_{\mu}/\partial x^{\varkappa} - \partial B^1{}_{\varkappa}/\partial x^{\mu} + i[B^1{}_{\varkappa}, B^1{}_{\mu}]$$

The following seven equations have an implicit sum over the four layers. The sums have no cross terms between layers since the respective generator matrices commute.

$$F_{SU(3)}{}^2{}_{\varkappa\mu} = \partial A_{SU(3)}{}^2{}_{\mu}/\partial x^{\varkappa} - \partial A_{SU(3)}{}^2{}_{\varkappa}/\partial x^{\mu} + ig_1[A_{SU(3)}{}^2{}_{\varkappa}, A_{SU(3)}{}^2{}_{\mu}] + ig_1[A_{SU(3)}{}^1{}_{\varkappa}, A_{SU(3)}{}^2{}_{\mu}] +$$
$$+ ig_1[A_{SU(3)}{}^2{}_{\varkappa}, A_{SU(3)}{}^1{}_{\mu}]$$

$$F_{W}{}^2{}_{\varkappa\mu} = \partial W^2{}_{\mu}/\partial x^{\varkappa} - \partial W^2{}_{\varkappa}/\partial x^{\mu} + ig_2[W^2{}_{\varkappa}, W^2{}_{\mu}] + ig_2[W^1{}_{\varkappa}, W^2{}_{\mu}] + ig_2[W^2{}_{\varkappa}, W^1{}_{\mu}]$$

$$F_{E}{}^2{}_{\varkappa\mu} = \partial A_{E}{}^2{}_{\mu}/\partial x^{\varkappa} - \partial A_{E}{}^2{}_{\varkappa}/\partial x^{\mu}$$

$$F_{DW}{}^2{}_{\varkappa\mu} = \partial W_{D}{}^2{}_{\mu}/\partial x^{\varkappa} - \partial W_{D}{}^2{}_{\varkappa}/\partial x^{\mu} + ig_4[W_{D}{}^2{}_{\varkappa}, W_{D}{}^2{}_{\mu}] + ig_4[W_{D}{}^1{}_{\varkappa}, W_{D}{}^2{}_{\mu}] +$$
$$+ ig_4[W_{D}{}^2{}_{\varkappa}, W_{D}{}^1{}_{\mu}]$$

$$F_{DE}{}^2{}_{\varkappa\mu} = \partial A_{DE}{}^2{}_{\mu}/\partial x^{\varkappa} - \partial A_{DE}{}^2{}_{\varkappa}/\partial x^{\mu}$$

$$F_{U}{}^2{}_{\varkappa\mu} = \partial U^2{}_{\mu}/\partial x^{\varkappa} - \partial U^2{}_{\varkappa}/\partial x^{\mu} + ig_6[U^2{}_{\varkappa}, U^2{}_{\mu}] + ig_6[U^1{}_{\varkappa}, U^2{}_{\mu}] + ig_6[U^2{}_{\varkappa}, U^1{}_{\mu}]$$

$$F_{V}{}^2{}_{\varkappa\mu} = \partial V^2{}_{\mu}/\partial x^{\varkappa} - \partial V^2{}_{\varkappa}/\partial x^{\mu} + ig_7[V^2{}_{\varkappa}, V^2{}_{\mu}] + ig_7[V^1{}_{\varkappa}, V^2{}_{\mu}] + ig_7[V^2{}_{\varkappa}, V^1{}_{\mu}]$$

The following three equations have *NO* implicit sum. They are not layer dependent.

$$F_{S}{}^2{}_{\varkappa\mu} = \partial A_{S}{}^2{}_{\mu}/\partial x^{\varkappa} - \partial A_{S}{}^2{}_{\varkappa}/\partial x^{\mu} + ig_8[A_{S}{}^2{}_{\varkappa}, A_{S}{}^2{}_{\mu}] + ig_8[A_{S}{}^1{}_{\varkappa}, A_{S}{}^2{}_{\mu}] + ig_8[A_{S}{}^2{}_{\varkappa}, A_{S}{}^1{}_{\mu}]$$

$$F_{\Theta}{}^2{}_{\varkappa\mu} = \partial A_{\Theta}{}^2{}_{\mu}/\partial x^{\varkappa} - \partial A_{\Theta}{}^2{}_{\varkappa}/\partial x^{\mu} + ig_9[A_{\Theta}{}^2{}_{\varkappa}, A_{\Theta}{}^2{}_{\mu}] + ig_9[A_{\Theta}{}^1{}_{\varkappa}, A_{\Theta}{}^2{}_{\mu}] + ig_9[A_{\Theta}{}^2{}_{\varkappa}, A_{\Theta}{}^1{}_{\mu}]$$

$$F_{B}{}^2{}_{\varkappa\mu} = \partial B^2{}_{\mu}/\partial x^{\varkappa} - \partial B^2{}_{\varkappa}/\partial x^{\mu} + i[B^1{}_{\mu}, B^2{}_{\varkappa}] + i[B^1{}_{\mu}, B^2{}_{\varkappa}] + i[B^2{}_{\mu}, B^1{}_{\varkappa}]$$

Note that $R'^{\beta}{}_{\sigma\nu\mu}$ factorizes into $[U(1)\otimes SU(2)\otimes U(1)\otimes SU(2)\otimes SU(3)\otimes U(4)\otimes U(4)]^4\otimes U(192)$ parts and a Riemann-Christoffel Gravitational curvature tensor part. For later use in defining a lagrangian we define (again with implicit sums over layers for the seven interactions in eq. 2.21)

$$R'^{\beta}{}_{\sigma\nu\mu} = R'_{E}{}^{1\beta}{}_{\sigma\nu\mu} + R'_{E}{}^{2\beta}{}_{\sigma\nu\mu} + R'_{SU(2)}{}^{1\beta}{}_{\sigma\nu\mu} + R'_{SU(2)}{}^{2\beta}{}_{\sigma\nu\mu} + R'_{DE}{}^{1\beta}{}_{\sigma\nu\mu} + R'_{DE}{}^{2\beta}{}_{\sigma\nu\mu} + R'_{DSU(2)}{}^{1\beta}{}_{\sigma\nu\mu} +$$
$$+ R'_{DSU(2)}{}^{2\beta}{}_{\sigma\nu\mu} + R'_{SU(3)}{}^{1\beta}{}_{\sigma\nu\mu} + R'_{SU(3)}{}^{2\beta}{}_{\sigma\nu\mu} + R'_{U}{}^{1\beta}{}_{\sigma\nu\mu} + R'_{U}{}^{2\beta}{}_{\sigma\nu\mu} + R'_{V}{}^{1\beta}{}_{\sigma\nu\mu} + R'_{V}{}^{2\beta}{}_{\sigma\nu\mu} +$$

$$+ \; R'_S{}^{1\beta}{}_{\sigma\nu\mu} + R'_S{}^{2\beta}{}_{\sigma\nu\mu} + R'_\Theta{}^{1\beta}{}_{\sigma\nu\mu} + R'_\Theta{}^{2\beta}{}_{\sigma\nu\mu} + R'_B{}^{1\beta}{}_{\sigma\nu\mu} + R'_B{}^{2\beta}{}_{\sigma\nu\mu} + R^{1\beta}{}_{\sigma\nu\mu} + R^{2\beta}{}_{\sigma\nu\mu}$$

$$(2.34)$$

where (with implicit sums over layers for the seven interactions in eq. 2.21)

$$R'_E{}^{1\beta}{}_{\sigma\nu\mu} = ig^\beta{}_\sigma F_E{}^1{}_{\nu\mu}$$
$$R'_E{}^{2\beta}{}_{\sigma\nu\mu} = ig^\beta{}_\sigma F_{DE}{}^2{}_{\nu\mu}$$

$$R'_{DE}{}^{1\beta}{}_{\sigma\nu\mu} = ig^\beta{}_\sigma F_E{}^1{}_{\nu\mu}$$
$$R'_{DE}{}^{2\beta}{}_{\sigma\nu\mu} = ig^\beta{}_\sigma F_{DE}{}^2{}_{\nu\mu}$$

$$R'_{SU(2)}{}^{1\beta}{}_{\sigma\nu\mu} = ig^\beta{}_\sigma F_W{}^1{}_{\nu\mu}$$
$$R'_{SU(2)}{}^{2\beta}{}_{\sigma\nu\mu} = ig^\beta{}_\sigma F_{DW}{}^2{}_{\nu\mu}$$

$$R'_{DSU(2)}{}^{1\beta}{}_{\sigma\nu\mu} = ig^\beta{}_\sigma F_W{}^1{}_{\nu\mu}$$
$$R'_{DSU(2)}{}^{2\beta}{}_{\sigma\nu\mu} = ig^\beta{}_\sigma F_{DW}{}^2{}_{\nu\mu}$$

$$R'_{SU(3)}{}^{1\beta}{}_{\sigma\nu\mu} = ig^\beta{}_\sigma F_{SU(3)}{}^1{}_{\nu\mu}$$
$$R'_{SU(3)}{}^{2\beta}{}_{\sigma\nu\mu} = ig^\beta{}_\sigma F_{SU(3)}{}^2{}_{\nu\mu}$$

$$R'_U{}^{1\beta}{}_{\sigma\nu\mu} = ig^\beta{}_\sigma F_U{}^1{}_{\nu\mu}$$
$$R'_U{}^{2\beta}{}_{\sigma\nu\mu} = ig^\beta{}_\sigma F_U{}^2{}_{\nu\mu}$$
$$R'_V{}^{1\beta}{}_{\sigma\nu\mu} = ig^\beta{}_\sigma F_V{}^1{}_{\nu\mu}$$
$$R'_V{}^{2\beta}{}_{\sigma\nu\mu} = ig^\beta{}_\sigma F_V{}^2{}_{\nu\mu}$$

$$R'_S{}^{1\beta}{}_{\sigma\nu\mu} = ig^\beta{}_\sigma F_S{}^1{}_{\nu\mu}$$
$$R'_S{}^{2\beta}{}_{\sigma\nu\mu} = ig^\beta{}_\sigma F_S{}^2{}_{\nu\mu}$$
$$R'_\Theta{}^{1\beta}{}_{\sigma\nu\mu} = ig^\beta{}_\sigma F_\Theta{}^1{}_{\nu\mu}$$
$$R'_\Theta{}^{2\beta}{}_{\sigma\nu\mu} = ig^\beta{}_\sigma F_\Theta{}^2{}_{\nu\mu}$$

$$R'_B{}^{1\beta}{}_{\sigma\nu\mu} = ig^\beta{}_\sigma B^1{}_{\nu\mu}$$
$$R'_B{}^{2\beta}{}_{\sigma\nu\mu} = ig^\beta{}_\sigma B^2{}_{\nu\mu}$$

The total Ricci tensor is

$$R'_{\sigma\mu} = R'{}^\beta{}_{\sigma\beta\mu} \qquad\qquad (2.35)$$

$$= iF_E^1{}_{\sigma\mu} + iF_E^2{}_{\sigma\mu} + iF_W^1{}_{\sigma\mu} + iF_W^2{}_{\sigma\mu} + iF_{DE}^1{}_{\sigma\mu} + iF_{DE}^2{}_{\sigma\mu} + iF_{DW}^1{}_{\sigma\mu} + iF_{DW}^2{}_{\sigma\mu} + iF_{SU(3)}^1{}_{\sigma\mu} + iF_{SU(3)}^2{}_{\sigma\mu} +$$
$$+ iF_U^1{}_{\sigma\mu} + iF_U^2{}_{\sigma\mu} + iF_V^1{}_{\sigma\mu} + iF_V^2{}_{\sigma\mu} + iF_S^1{}_{\sigma\mu} + iF_S^2{}_{\sigma\mu} + iF_\Theta^1{}_{\sigma\mu} + iF_\Theta^2{}_{\sigma\mu} + iB^1{}_{\sigma\mu} + iB^2{}_{\sigma\mu} +$$
$$+ \partial_\mu H^{1\beta}{}_{\sigma\beta} - \partial_\beta H^{1\beta}{}_{\sigma\mu} + H^{1\gamma}{}_{\beta\sigma} H^{1\beta}{}_{\gamma\mu} - H^{1\gamma}{}_{\mu\sigma} H^{1\beta}{}_{\gamma\beta} +$$
$$+ \partial_\mu H^{2\beta}{}_{\sigma\beta} - \partial_\beta H^{2\beta}{}_{\sigma\mu} + H^{2\gamma}{}_{\beta\sigma} H^{2\beta}{}_{\gamma\mu} - H^{2\gamma}{}_{\mu\sigma} H^{2\beta}{}_{\gamma\beta} + H^{1\gamma}{}_{\beta\sigma} H^{2\beta}{}_{\gamma\mu} - H^{1\gamma}{}_{\mu\sigma} H^{2\beta}{}_{\gamma\beta} + H^{2\gamma}{}_{\beta\sigma} H^{1\beta}{}_{\gamma\mu} - H^{2\gamma}{}_{\mu\sigma} H^{1\beta}{}_{\gamma\beta}$$

$$= R'_E{}^1{}_{\sigma\mu} + R'_E{}^2{}_{\sigma\mu} + R'_{SU(2)}{}^1{}_{\sigma\mu} + R'_{SU(2)}{}^2{}_{\sigma\mu} + R'_{DE}{}^1{}_{\sigma\mu} + R'_{DE}{}^2{}_{\sigma\mu} + R'_{DSU(2)}{}^1{}_{\sigma\mu} + R'_{DSU(2)}{}^2{}_{\sigma\mu} + R'_{SU(3)}{}^1{}_{\sigma\mu} +$$
$$+ R'_{SU(3)}{}^2{}_{\sigma\mu} + R'_U{}^1{}_{\sigma\mu} + R'_U{}^2{}_{\sigma\mu} + R'_V{}^1{}_{\sigma\mu} + R'_V{}^2{}_{\sigma\mu} + R'_S{}^1{}_{\sigma\mu} + R'_S{}^2{}_{\sigma\mu} + R'_\Theta{}^1{}_{\sigma\mu} + R'_\Theta{}^2{}_{\sigma\mu} + R'_B{}^{1\beta}{}_{\sigma\beta\mu} +$$
$$+ R'_B{}^{2\beta}{}_{\sigma\beta\mu} + R^1{}_{\sigma\mu} + R^2{}_{\sigma\mu}$$
$$= R'^1{}_{\sigma\mu} + R'^2{}_{\sigma\mu} \tag{2.36}$$

where (with implicit sums over layers for the seven interactions in eq. 2.21. Again there are no cross terms between layers.)

$$R'^1{}_{\sigma\mu} = R'_E{}^1{}_{\sigma\mu} + R'_{SU(2)}{}^1{}_{\sigma\mu} + R'_{DE}{}^1{}_{\sigma\mu} + R'_{DSU(2)}{}^1{}_{\sigma\mu} + R'_{SU(3)}{}^1{}_{\sigma\mu} + R'_U{}^1{}_{\sigma\mu} + R'_V{}^1{}_{\sigma\mu} + R'_S{}^1{}_{\sigma\mu} +$$
$$+ R'_\Theta{}^1{}_{\sigma\mu} + R'_B{}^{1\beta}{}_{\sigma\beta\mu} + R^1{}_{\sigma\mu} \tag{2.37}$$

$$R'^2{}_{\sigma\mu} = R'_E{}^2{}_{\sigma\mu} + R'_{SU(2)}{}^2{}_{\sigma\mu} + R'_{DE}{}^2{}_{\sigma\mu} + R'_{DSU(2)}{}^2{}_{\sigma\mu} + R'_{SU(3)}{}^2{}_{\sigma\mu} + R'_U{}^2{}_{\sigma\mu} + R'_V{}^2{}_{\sigma\mu} + R'_S{}^2{}_{\sigma\mu} +$$
$$+ R'_\Theta{}^2{}_{\sigma\mu} + R'_B{}^{2\beta}{}_{\sigma\beta\mu} + R^2{}_{\sigma\mu}$$

with

$$R'_E{}^1{}_{\sigma\mu} = iF_E{}^1{}_{\sigma\mu}$$
$$R'_E{}^2{}_{\sigma\mu} = iF_E{}^2{}_{\sigma\mu}$$

$$R'_{SU(2)}{}^1{}_{\sigma\mu} = iF_W{}^1{}_{\sigma\mu}$$
$$R'_{SU(2)}{}^2{}_{\sigma\mu} = iF_W{}^2{}_{\sigma\mu}$$

$$R'_{DE}{}^1{}_{\sigma\mu} = iF_{DE}{}^1{}_{\sigma\mu}$$
$$R'_{DE}{}^2{}_{\sigma\mu} = iF_{DE}{}^2{}_{\sigma\mu}$$

$$R'_{DSU(2)}{}^1{}_{\sigma\mu} = iF_{DW}{}^1{}_{\sigma\mu}$$
$$R'_{DSU(2)}{}^2{}_{\sigma\mu} = iF_{DW}{}^2{}_{\sigma\mu}$$

$$R'_{SU(3)}{}^1{}_{\sigma\mu} = iF_{SU(3)}{}^1{}_{\sigma\mu}$$
$$R'_{SU(3)}{}^2{}_{\sigma\mu} = iF_{SU(3)}{}^2{}_{\sigma\mu}$$

$$R'_U{}^1{}_{\sigma\mu} = iF_U{}^1{}_{\sigma\mu}$$
$$R'_U{}^2{}_{\sigma\mu} = iF_U{}^2{}_{\sigma\mu}$$

$$R'^{1}_{V\,\sigma\mu} = iF^{1}_{V\,\sigma\mu}$$
$$R'^{2}_{V\,\sigma\mu} = iF^{2}_{V\,\sigma\mu}$$

$$R'^{1}_{S\,\sigma\mu} = iF^{1}_{S\,\sigma\mu}$$
$$R'^{2}_{S\,\sigma\mu} = iF^{2}_{S\,\sigma\mu}$$

$$R'^{1}_{\Theta\,\sigma\mu} = iF^{1}_{\Theta\,\sigma\mu}$$
$$R'^{2}_{\Theta\,\sigma\mu} = iF^{2}_{\Theta\,\sigma\mu}$$

$$R'^{1}_{B\,\sigma\mu} = iB^{1}_{\sigma\mu}$$
$$R'^{2}_{B\,\sigma\mu} = iB^{2}_{\sigma\mu}$$

with the further definition of $R''^{1}_{\sigma\mu}$ and $R''^{2}_{\sigma\mu}$:

$$R''^{1}_{\sigma\mu} = R'_{SU(3)}{}^{1}_{\sigma\mu} + R^{1}_{\sigma\mu} \qquad (2.38)$$
$$R''^{2}_{\sigma\mu} = R'_{SU(3)}{}^{2}_{\sigma\mu} + R^{2}_{\sigma\mu}$$

Eq. 2.37 is the Ricci tensor. An additional Ricci-like tensor is

$$H_{\sigma\mu} = H^{\beta}_{\sigma\beta\mu} \qquad (2.39)$$

The curvature scalar is

$$R' = g^{\sigma\mu}R'_{\sigma\mu} = + \partial^{\sigma}H^{1\beta}_{\sigma\beta} - \partial_{\beta}H^{1\beta}{}_{\sigma}{}^{\sigma} + H^{1\gamma}_{\beta\sigma}H^{1\beta}{}_{\gamma}{}^{\sigma} - H^{1\gamma}_{\mu\sigma}H^{1\beta}_{\gamma\beta} + \partial^{\sigma}H^{2\beta}_{\sigma\beta} - \partial_{\beta}H^{2\beta}{}_{\sigma}{}^{\sigma} +$$
$$+ H^{2\gamma}_{\beta\sigma}H^{2\beta}{}_{\gamma}{}^{\sigma} - H^{2\gamma\sigma}{}_{\sigma}H^{2\beta}_{\gamma\beta} + H^{1\gamma}_{\beta\sigma}H^{2\beta}{}_{\gamma}{}^{\sigma} - H^{1\gamma\sigma}{}_{\sigma}H^{2\beta}_{\gamma\beta} + H^{2\gamma}_{\beta\sigma}H^{1\beta}{}_{\gamma}{}^{\sigma} - H^{2\gamma\sigma}{}_{\sigma}H^{1\beta}_{\gamma\beta}$$

$$= g^{\sigma\mu}(R^{1\beta}_{\sigma\beta\mu} + R^{2\beta}_{\sigma\beta\mu}) \qquad (2.40)$$

2.11 SuperStandard Lagrangian

To construct the vector and gravitational bosonic sector we must separate the various parts of $R'^{1}_{\sigma\mu}$ and $R''^{1}_{\sigma\mu}$ into pieces for each layer. From eq. 2.37 we therefore define

$$^{a}R'^{1}_{\sigma\mu} = {}^{a}R'^{1}_{E\,\sigma\mu} + {}^{a}R'_{SU(2)}{}^{1}_{\sigma\mu} + {}^{a}R'_{DE}{}^{1}_{\sigma\mu} + {}^{a}R'_{DSU(2)}{}^{1}_{\sigma\mu} + {}^{a}R'_{SU(3)}{}^{1}_{\sigma\mu} + {}^{a}R'^{1}_{U\,\sigma\mu} + {}^{a}R'^{1}_{V\,\sigma\mu}$$
$$^{grav}R'^{1}_{\sigma\mu} = + R'^{1}_{S\,\sigma\mu} + R'^{1}_{\Theta\,\sigma\mu} + R'_{B}{}^{1\beta}_{\sigma\beta\mu} + R^{1}_{\sigma\mu} \qquad (2.41)$$

and

$$^{a}R'^{2}_{\sigma\mu} = {}^{a}R'^{2}_{E\,\sigma\mu} + {}^{a}R'_{SU(2)}{}^{2}_{\sigma\mu} + {}^{a}R'_{DE}{}^{2}_{\sigma\mu} + {}^{a}R'_{DSU(2)}{}^{2}_{\sigma\mu} + {}^{a}R'_{SU(3)}{}^{2}_{\sigma\mu} + {}^{a}R'^{2}_{U\,\sigma\mu} + {}^{a}R'^{2}_{V\,\sigma\mu}$$
$$^{grav}R'^{2}_{\sigma\mu} = R'^{2}_{S\,\sigma\mu} + R'^{2}_{\Theta\,\sigma\mu} + R'_{B}{}^{2\beta}_{\sigma\beta\mu} + R^{2}_{\sigma\mu} \qquad (2.42)$$

using the fields and coupling constants of each layer to define the expressions for each $^{a}R'..._{\sigma\mu}^{i}$ in terms of the corresponding $^{a}F..._{\sigma\mu}^{i}$ in a similar fashion to eq. 2.37 above.

The boson part of the lagrangian of the Unified SuperStandard Model (with the fermion sector, Higgs sector and the Faddeev-Popov terms gauge sector not displayed here[40]) is:

$$\mathcal{L} = \text{Tr } \sqrt{g}[\Sigma_a(MD_v{}^aR''^1{}_{\sigma\mu}D^{va}R''^{2\sigma\mu} + a^aR'^1{}_{\sigma\mu}{}^aR'^{2\sigma\mu}) + MD_v{}^{grav}R''^1{}_{\sigma\mu}D^{vgrav}R''^{2\sigma\mu} + a^{grav}R'^1{}_{\sigma\mu}{}^{grav}R'^{2\sigma\mu} +$$
$$+ bR' + cg^{\sigma\mu}g^2{}_{\sigma\mu} + c'g^{2\sigma\mu}g^2{}_{\sigma\mu} - d\Sigma_a{}^aA_{SU(3)}{}^2{}_\mu{}^aA_{SU(3)}{}^{2\mu}] \tag{2.43}$$

where M, a, b, c, c', and d are constants and a is the layer index. Note $R''^i{}_{\sigma\mu}$ for i = 1, 2 only contains Strong Interaction and Gravitation terms. Eq. 2.43 maintains the independence of each layer's particle interactions. The terms with $^{grav}R''^i{}_{\sigma\mu}$ and $R's^i{}_{\sigma\mu}$ for i = 1, 2 are for universal gravitation and Θ-symmetry, and apply equally to all layers. Coupling constants and masses are layer dependent.

This higher derivative lagrangian maintains the locality of the theory but does entail a modest, yet still canonical, modification in the derivation of the Euler-Lagrange equations of motion. It also requires the use of principal value propagators rather than ordinary Feynman propagators for gluon and graviton interactions. Thus the Strong Interaction sector, and the Gravitation sector are Action-at-a-Distance theories that are similar in spirit to Wheeler-Feynman Electrodynamics. The two U(1) Electromagnetic sectors, the Generation group U(4) gauge field sector, the Layer group U(4) gauge field sector, the two SU2) Weak sectors, the U(4) A_S gauge field sector, the spinor connection sector, and the Θ-interaction sector may, or may not, be Action-at-a-Distance fields. They are not constrained to be Action-at-a-Distance by the present considerations.

Eq. 2.43 can therefore be expressed as:[41]

$$\mathcal{L} = \mathcal{L}_E + \mathcal{L}_{SU(2)} + \mathcal{L}_{DE} + \mathcal{L}_{DSU(2)} + \mathcal{L}_{SU(3)} + \mathcal{L}_U + \mathcal{L}_V + \mathcal{L}_S + \mathcal{L}_\Theta + \mathcal{L}_G + \mathcal{L}_{int} \tag{2.43}$$

with the understanding that the first seven terms are sums of four independent layer terms of the specified interactions. The $\mathcal{L}_{int}$ terms also contain sums of layer terms. Traces of lagrangian terms are implicit. We can separate the lagrangian terms as follows

$$\mathcal{L}_E = \text{Tr } \Sigma_a\sqrt{g}\{a^aF_E^1{}_{\sigma\mu}{}^aF_E^{2\sigma\mu}\}$$
$$\mathcal{L}_{SU(2)} = \text{Tr } \Sigma_a\sqrt{g}[a^aF_W^1{}_{\sigma\mu}{}^aF_W^{2\sigma\mu}]$$
$$\mathcal{L}_{DE} = \text{Tr } \Sigma_a\sqrt{g}\{a^aF_{DE}^1{}_{\sigma\mu}{}^aF_{DE}^{2\sigma\mu}\}$$
$$\mathcal{L}_{DSU(2)} = \text{Tr } \Sigma_a\sqrt{g}[a^aF_{DW}^1{}_{\sigma\mu}{}^aF_{DW}^{2\sigma\mu}]$$
$$\mathcal{L}_{SU(3)} = \text{Tr } \Sigma_a\sqrt{g}\{M[\partial_v + i(^aA_{SU(3)}^1{}_v + {}^aA_{SU(3)}^2{}_v)]^aF_{SU(3)}^1{}_{\sigma\mu}[\partial^v + i(^aA_{SU(3)}^{1v} + {}^aA_{SU(3)}^{2v})]^aF_{SU(3)}^{2\sigma\mu} +$$

[40] See volume I for these items.
[41] We only consider the gauge field lagrangian terms.

$$+ {}^a F_{SU(3)}{}^1_{\sigma\mu}\, {}^a F_{SU(3)}{}^{2\sigma\mu} - d^a A_{SU(3)}{}^2_\mu\, {}^a A_{SU(3)}{}^{2\mu}\}$$

$$\mathcal{L}_U = Tr\ \Sigma_a \sqrt{g}[a^a F_U{}^1_{\sigma\mu}\, {}^a F_U{}^{2\sigma\mu}]$$

$$\mathcal{L}_V = Tr\ \Sigma_a \sqrt{g}[a^a F_V{}^1_{\sigma\mu}\, {}^a F_V{}^{2\sigma\mu}]$$

$$\mathcal{L}_S = Tr\ \sqrt{g}[a F_S{}^1_{\sigma\mu} F_S{}^{2\sigma\mu}]$$

$$\mathcal{L}_\Theta = Tr\ \sqrt{g}[a F_\Theta{}^1_{\sigma\mu} F_\Theta{}^{2\sigma\mu}]$$

$$\mathcal{L}_G = Tr\ \sqrt{g}[MD_v R^1_{\sigma\mu} D^v R^{2\sigma\mu} + a R^1_{\sigma\mu} R^{2\sigma\mu} + b g^{\sigma\mu}(R^{1\beta}{}_{\sigma\beta\mu} + R^{2\beta}{}_{\sigma\beta\mu}) + c g^{\sigma\mu} g^2_{\sigma\mu} + c' g^{2\sigma\mu} g^2_{\sigma\mu}]$$

$$= Tr\ \sqrt{g}[MD_v R^1_{\sigma\mu} D^v R^{2\sigma\mu} + a R^1_{\sigma\mu} R^{2\sigma\mu} + bH + c g^{\sigma\mu} g^2_{\sigma\mu} + c' g^{2\sigma\mu} g^2_{\sigma\mu}]$$

where $R^1_{\sigma\mu}$ and $R^1_{\sigma\mu}$ are given in eqs. 2.31 and 2.32. The remaining terms are interactions between the vector boson fields that do not appear to have been considered before our work:

$$\mathcal{L}_{int} = \mathcal{L} - (\mathcal{L}_E + \mathcal{L}_{SU(2)} + \mathcal{L}_{DE} + \mathcal{L}_{DSU(2)} + \mathcal{L}_{SU(3)} + \mathcal{L}_U + \mathcal{L}_V + \mathcal{L}_S + \mathcal{L}_\Theta + \mathcal{L}_G)$$

$\mathcal{L}_{SU(3)}$, $\mathcal{L}_{SU(2)}$, $\mathcal{L}_E$, $\mathcal{L}_{DE}$, $\mathcal{L}_{DSU(2)}$, $\mathcal{L}_U$, $\mathcal{L}_V$, $\mathcal{L}_S$, $\mathcal{L}_\Theta$, and parts of $\mathcal{L}_{int}$ are the dominant interactions within hadrons, and $\mathcal{L}_G$, $\mathcal{L}_E$ and parts of $\mathcal{L}_{int}$ are the dominant interactions in space within the framework of this discussion.

The $D_v R^1_{\sigma\mu}$ and $D^v R^{2\sigma\mu}$ terms have the form:

$$D_v R^i_{\sigma\mu} = + \partial_v R^i_{\sigma\mu} - H^{1\beta}{}_{\sigma v} R^i_{\beta\mu} - H^{1\beta}{}_{v\mu} R^i_{\sigma\beta}$$

for i = 1, 2.

2.12 New SuperStandard Interactions

The above lagrangian terms can be expressed in the following manner:

$$\mathcal{L}_E = Tr\ \Sigma_a \sqrt{g}\{a^a F_E{}^1_{\sigma\mu}\, {}^a F_E{}^{2\sigma\mu}\}$$

$$\mathcal{L}_{SU(2)} = Tr\ \Sigma_a \sqrt{g}[a^a F_W{}^1_{\sigma\mu}\, {}^a F_W{}^{2\sigma\mu}]$$

$$\mathcal{L}_{DE} = Tr\ \Sigma_a \sqrt{g}\{a^a F_{DE}{}^1_{\sigma\mu}\, {}^a F_{DE}{}^{2\sigma\mu}\}$$

$$\mathcal{L}_{DSU(2)} = Tr\ \Sigma_a\, g[a^a F_{DW}{}^1_{\sigma\mu}\, {}^a F_{DW}{}^{2\sigma\mu}]$$

$$\mathcal{L}_{SU(3)} = Tr\ \Sigma_a \sqrt{g}\{M[\partial_v + i({}^a A_{SU(3)}{}^1_v + {}^a A_{SU(3)}{}^2_v)]\, {}^a F_{SU(3)}{}^1_{\sigma\mu}[\partial^v + i({}^a A_{SU(3)}{}^{1v} + {}^a A_{SU(3)}{}^{2v})]\, {}^a F_{SU(3)}{}^{2\sigma\mu} +$$

$$+ {}^a F_{SU(3)}{}^1_{\sigma\mu}\, {}^a F_{SU(3)}{}^{2\sigma\mu} - d^a A_{SU(3)}{}^2_\mu\, {}^a A_{SU(3)}{}^{2\mu}\}$$

$$\mathcal{L}_U = Tr\ \Sigma_a \sqrt{g}[a^a F_U{}^1_{\sigma\mu}\, {}^a F_U{}^{2\sigma\mu}]$$

$$\mathcal{L}_V = Tr\ \Sigma_a \sqrt{g}[a^a F_V{}^1_{\sigma\mu}\, {}^a F_V{}^{2\sigma\mu}]$$

$$\mathcal{L}_S = Tr\ \Sigma_a \sqrt{g}[a F_S{}^1_{\sigma\mu} F_S{}^{2\sigma\mu}]$$

$$\mathcal{L}_\Theta = Tr\ \sqrt{g}[a F_\Theta{}^1_{\sigma\mu} F_\Theta{}^{2\sigma\mu}]$$

$$\mathcal{L}_G = Tr\ \sqrt{g}[MD_v R^1_{\sigma\mu} D^v R^{2\sigma\mu} + a R^1_{\sigma\mu} R^{2\sigma\mu} + b g^{\sigma\mu}(R^{1\beta}{}_{\sigma\beta\mu} + R^{2\beta}{}_{\sigma\beta\mu}) + c g^{\sigma\mu} g^2_{\sigma\mu} + c' g^{2\sigma\mu} g^2_{\sigma\mu}]$$

$$= Tr\ \sqrt{g}[MD_v R^1_{\sigma\mu} D^v R^{2\sigma\mu} + a R^1_{\sigma\mu} R^{2\sigma\mu} + bH + c g^{\sigma\mu} g^2_{\sigma\mu} + c' g^{2\sigma\mu} g^2_{\sigma\mu}]$$

$$\mathcal{L}_{int} = \mathcal{L} - (\mathcal{L}_E + \mathcal{L}_{SU(2)} + \mathcal{L}_{DE} + \mathcal{L}_{DSU(2)} + \mathcal{L}_{SU(3)} + \mathcal{L}_U + \mathcal{L}_V + \mathcal{L}_S + \mathcal{L}_\Theta + \mathcal{L}_G)$$

Thus $\mathcal{L}_{SU(3)}$, $\mathcal{L}_{SU(2)}$, $\mathcal{L}_E$, $\mathcal{L}_{DE}$, $\mathcal{L}_{DSU(2)}$, $\mathcal{L}_U$, $\mathcal{L}_V$, $\mathcal{L}_S$, $\mathcal{L}_\Theta$, and parts of $\mathcal{L}_{int}$ are the dominant interactions within hadrons, and $\mathcal{L}_G$, $\mathcal{L}_E$ and parts of $\mathcal{L}_{int}$ are the dominant interactions in space within the framework of this discussion. *The terms of $\mathcal{L}_{int}$ have 'new' interactions between gauge fields, some of which are described in chapters 28 through 30 of volume I. These interactions are not in the conventional Standard Model.*

Chapter 28[42] show are lagrangian creates a gravitational field with three distinct parts: at solar system distances it gives a V(r) = G/r potential, at galactic distances it gives

$$V(r) \cong -[G + 1/(96\pi a)]/r + \tfrac{1}{2}G\cos(m_G r)/r \qquad (28.8)$$

and at inter-galactic distances it gives a weaker gravitational potential

$$V(r) \cong -[G + 1/(96\pi a)]/r \qquad (28.13)$$

Chapter 29 shows our theory gives a linear quark potential plus a 1/r term that corresponds to the Charmonium potential used to calculate the Charmonium particle spectrum by the Cornell group:

$$V(r) = -2f^2/r + f^2\lambda^2 r \qquad (29.6)$$

2.13 Other Effects of New Interactions Between Boson Interactions

There are numerous other effects embedded in our lagrangian that can be called vector boson – vector boson interactions. They specify new *three and four vector boson vertices* for the SU(3) and gravitational interactions. They are generated by the $M\Sigma_a D_v{}^a R''^1{}_{\sigma\mu} D^{va} R''^{2\sigma\mu}$ term. Below we give some examples *from volume I.*

2.13.1 Missing Nucleon Spin Puzzle

The estimates of nucleon spin that are obtained from parton analyses of deep inelastic electron – nucleon interactions are woefully short of the spin expected in quark models of nucleons. The missing spin has been attributed to a number of causes. However the Missing Spin Puzzle remains.

From eq. 27.3 - 27.4 of volume I it is clear that there are important new interaction terms between the Electromagnetic and Strong interaction fields. After taking traces we find

$$\mathcal{L}_{intEM-S} = -\mathrm{Tr}\ iM\{(A_E{}^1{}_v + A_E{}^2{}_v)\ F_{SU(3)}{}^1{}_{\sigma\mu} D^v F_{SU(3)}{}^{2\sigma\mu} + iD_v F_{SU(3)}{}^1{}_{\sigma\mu}(A_E{}^{1v} + A_E{}^{2v}) F_{SU(3)}{}^{2\sigma\mu}\} \qquad (30.1)$$

[42] Equation numbers in the remainder of this section and 2.13 are from volume I.

$\mathcal{L}_{intEM\text{-}S}$ generates a combined photon-gluon vertex insertion in gluon interactions between quarks within a hadron. Figs. 30.1 and 30.2 show two simple possible vertex insertions in a gluon line.

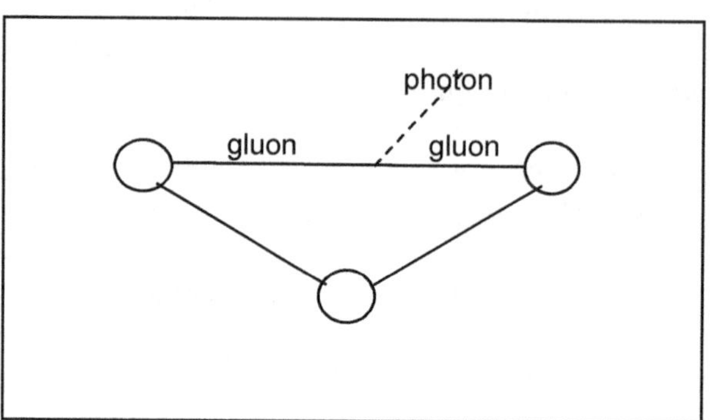

Figure 30.1. A single 'outgoing' photon vertex insertion in a gluon line for a three quark nucleon. Only single gluon lines between the three quarks are displayed.

The gluon line, by itself, has $1/k^4$ and $1/k^2$ momentum space propagator terms. *The insertion of the vertex in Fig.30.1 generated by $\mathcal{L}_{intEM\text{-}S}$ yields a combined momentum factor of $k^3(k^4k^4)^{-1} = k^{-5}$ which would make it (summed over all gluon lines) a significant contribution to the proton spin determination in deep inelastic electron-nucleon scattering.*[43] *The insertion of the vertex in Fig. 30.2 generated by $\mathcal{L}_{intEM\text{-}S}$ yields a combined momentum factor of $k^2(k^4k^4)^{-1} = k^{-6}$ which may have a less significant effect.*

[43] See C. A. Aidala, S. D. Bass, D. Hasch, and G. K. Mallot, arXiv: 1209.2803v2 (2013) and references therein for a review of the 'missing' nucleon spin puzzle.

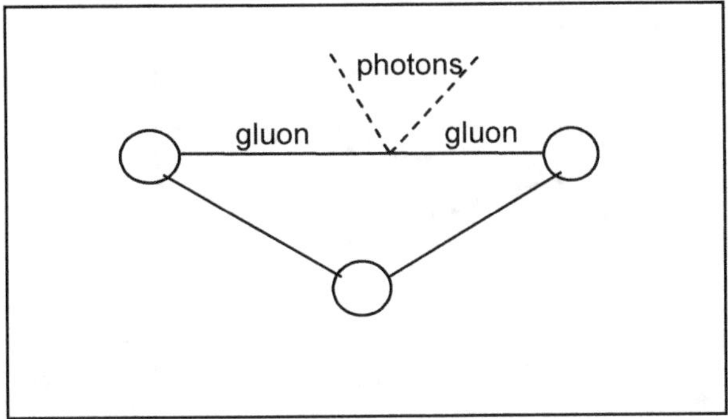

Figure 30.2. An 'outgoing' two photon vertex insertion in a gluon line for a three quark nucleon. Only single gluon lines between the three quarks are displayed.

Thus our unified theory may solve the nucleon spin puzzle. The interactions in Figs. 30.1 and 30.2 introduce a new direct connection between photons and spin one gluons. Thus their contributions to the summations of proton spin interactions in parton models may account for the 'missing' two-thirds of proton spin. Our unified theory has a new gluon-photon interaction that is not found in the conventional Standard Model.

2.13.2 Discrepancies between Proton Radius Measurements

Recently experiments have confirmed that the radius of a proton in a muonic hydrogen atom is smaller than the proton radius measured in a conventional hydrogen atom composed of a proton encircled by an electron. The lagrangian in eq. 27.1 indicates that there are direct interactions between photons and gluon gauge fields such as

$$A_{E\ \nu}^{\ 1}A_E^{\ 2\nu}A_{SU(3)}^{\ \ \ \ 1\mu}A_{SU(3)\ \mu}^{\ \ \ \ 2}$$

These interactions result in Feynman diagrams that modify the electromagnetic field between a proton and a circling muon or electron. Fig. 30.3 shows simple forms of this interaction for a muon and a quark within a proton.

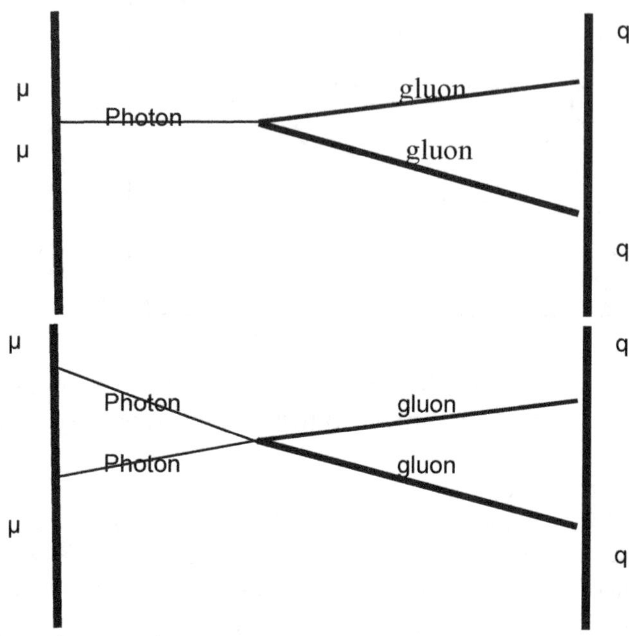

Figure 2.7. Photon-gluon interactions between a muon and quark (within a proton).

Our lagrangian also indicates that there are direct interactions between gluons and Generation group gauge fields of the form

$$\mathbf{A}_{SU(3)}{}^{\mu}\mathbf{A}_{SU(3)\mu}U_{\nu}U^{i\nu}$$

These interactions result in Feynman diagrams that modify the energy layers of a circling muon. Fig. 30.3 shows simple forms of this interaction for a muon and a quark within a proton.

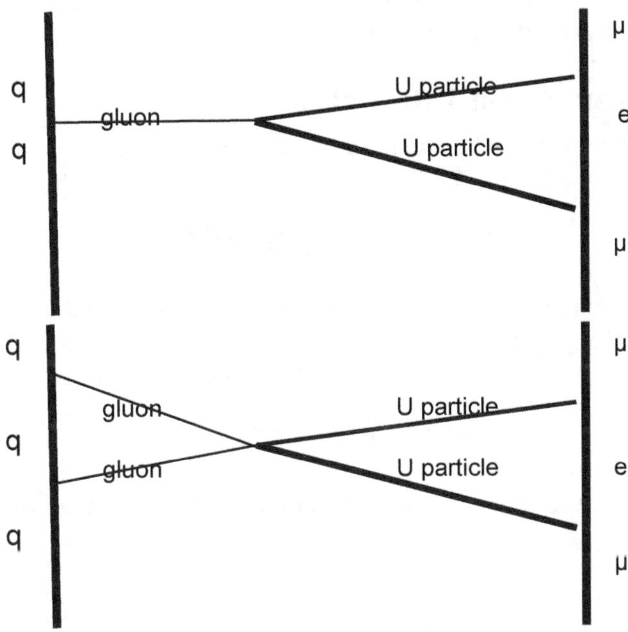

Figure 30.3. Gluon - U Generation group gauge particle interactions between a muon and quark (within a proton).

The net effect of these additional interactions, which are not in the Standard Model, is to increase the force between the quark and muon beyond what one would expect using the electromagnetic potential, from the usual

$$V = -e^2/r$$

to

$$V' = -(e+\delta)^2/r \qquad (30.2)$$
$$= -e^2/r'$$

where δ is small compared to e, and where

$$r' = r[e/(e+\delta)]^2 < r$$

giving an apparently smaller radial distance than the actual muon radial distance. This change would cause an energy shift to a lower value (more negative) in the muonic energy layers and, consequently, the proton radius would appear smaller for muonic atoms – as experiment confirms.[44]

[44] Fig. 30.3 is not of importance for 'normal' hydrogen since the electron is so much lighter than the muon.

Thus the new 'interactions between interactions' of our unified theory have significant effects that may be already qualitatively verified by experiment.

2.14 More Vector Boson Interactions

Additional interactions between gluons and other vector bosons exist in each layer for $\mathbf{W}^{i\mu}(x)$, $\mathbf{W_D}^{i\mu}(x)$, $\mathbf{A_{DE}}^{i\mu}(x)$, $\mathbf{U}^{i\mu}(x)$, and $\mathbf{V}^{i\mu}(x)$. These interactions are due to the

$$M\Sigma_a D_v {}^a R''^1{}_{\sigma\mu} D^v {}^a R''^{2\sigma\mu}$$

term. Because of the presumed weakness of the coupling constants of these interactions they may not be of experimental significance.

There are also gravitation-vector boson interaction due to the

$$M D_v {}^{grav}R''^1{}_{\sigma\mu} D^v {}^{grav}R''^{2\sigma\mu}$$

term that could be of importance in unusal cosmological phenomena such as neutron and quark stars.

3. Supersymmetry and the Unified SuperStandard Model

3.1 Initial Goals and Successes of SuperSymmetry

The initial hope[45] of SuperSymmetry was to unite bosons and fermions in one supermultiplet. This hope could not be realized because fermions lie in the fundamental representation of internal symmetry groups while vector bosons lie in the adjoint representation of internal symmetry groups. This dilemma was rectified by the introduction of new bosons and fermions to create two supermultiplets. The introduced particles had names such as squarks, sleptons, Gauginos, Higgsinos, and Goldstinos.

Supersymmetry theory was extended to included SuperSymmetric coordinates and Supersymmetry field theory was developed.[46] Perhaps the major benefit of SuperSymetry Field Theory was the proof that its perturbation theory was finite to all orders. This was a signal success. However our Two-Tier approach to Quantum Field Theory also yields a perturbation theory that is finite to all orders in any number of dimensions and for any polynomial lagrangian. Thus there is a greater alternative to SuperSymmetric Field Theory that works.

One must also recognize that SuperSymmetric coordinates have not been observed in nature. Consequently they may be an artificial construct that might be acceptable IF an overwhelming reason for their existence were found.

3.2 Unified SuperStandard Model Form of SuperSymmetry

Our new Unified SuperStandard Model has equal numbers of fermions and vector bosons. The vector bosons are directly related to the form of the fermion particles' structure – the Fermion Periodic Table. There are 192 fermions and 192 vector bosons.[47] The 192 fermions form a <u>192</u> representation of the Θ-Symmetry group, and the 192 vector bosons form another <u>192</u> representation of the U(192) Θ-Symmetry group.

[45] Myazawa, H., Prog. Theor. Phys. **36**, 1266 (1966); Phys. Rev. **170**, 1586 (1968); Gol'fand, Yu. A. and Likhtman, E. P., Sov. Phys.: JETP Lett. **13**, 323 (1971).

[46] Neveu, A. and Schwarz, J. H., Nucl Phys. **B31**, 86 (1971); Ramond, P., Phys. Rev. **D3**, 2415 (1971); Gervais, J. L. and Sakita, B., Nucl. Phys. **B34**, 632 (1971); Volkov, D. V. and Akulov, V. P., JETP Lett. **16**, 438 (1972); Wess, J. and Zumino, B., Nucl. Phys. **B70**, 34 (1974); and references therein.

[47] There are also Higgs bosons and Θ-Symmetry vector bosons. These vector bosons do not affect the number or form of the Fermion Periodic Table and thus can be viewed as a separate issue. In fact we will show the Θ-Symmetry group plays a role in establishing a SuperSymmetry formalism in the Unified SuperStandard Model.

3.2.1 The Θ-Symmetry Group

The U(192) Θ-Symmetry group has 192^2 gauge fields and generators. Elements of the group algebra form a closed set under commutation. The fundamental representation has 192 dimensions.

3.2.2 Fermionic Operators

One can define two sets[48] of 192^2 fermionic operators $\{A_{\Theta F}{}^{\mu}{}_{nm}(x)\}$ and $\{A_{\Theta B}{}^{\mu}{}_{nm}(x)\}$ that transform fermions into bosons between the <u>192</u> Θ-Symmetry fermion and boson representations; and that also transform bosons into fermions between the <u>192</u> Θ-Symmetry fermion and boson representations respectively:

$$A_{\Theta F}{}^{\mu}{}_{nm}(x)\psi_m(x) \rightarrow A^{\mu}{}_n$$
$$A_{\Theta B}{}^{\mu}{}_{nm}(x)A_{\mu m}(x) \rightarrow \psi_n \tag{3.1}$$

where n = 1, ... , 192 and m = 1, ... , 192 label multiplet fields irrespective of internal symmetries. Each n and m pair label one of 192^2 fermionic operators.

The fermionic operators are closed under anticommutation. One representation of these operators in terms of the fermionic and bosonic fields is

$$A_{\Theta F}{}^{\mu}{}_{nm}(x) = A^{\mu}{}_n(x)\overline{\psi}_m(x) \tag{3.2}$$
$$A_{\Theta B}{}^{\mu}{}_{nm}(x) = \psi_n(x)A^{\mu}{}_m(x)^{\dagger} \tag{3.3}$$

where n and m are labels ranging from 1, ... 192 for fermions and bosons, $A^{\mu}{}_n(x)$ is the n^{th} of the 192 vector fields, $\psi_m(x)$ is the m^{th} of the 192 fermion fields in eq. 3.2; and $A^{\mu}{}_m(x)^{\dagger}$ is the hermitean conjugate of the m^{th} vector field of the 192 vector fields and $\psi_n(x)$ is the n^{th} fermion field of the 192 fermion fields in eq. 3.3. Spatial integrals of eqs. 3.2 and 3.3 have simpler anticommutation relations.

Elements of the Θ-Symmetry group can be used to map the fermionic operators to different forms. For example if U_{Θ} is an element of the Θ-Symmetry group we can redefine fermion operators as a different map between fermions and bosons with expresions like

$$A'_{\Theta F}{}^{\mu}{}_{nm} = U_{\Theta}A_{\Theta F}{}^{\mu}{}_{nm}U_{\Theta}^{-1} \tag{3.4}$$
$$A'_{\Theta B}{}^{\mu}{}_{nm} = U_{\Theta}A_{\Theta B}{}^{\mu}{}_{nm}U_{\Theta}^{-1} \tag{3.5}$$

Thus one can SuperSymmetric map between the fermion and vector boson multiplets. We have not displayed internal symmetry indices on the fields. Clearly the difference between the internal symmetry fundamental representation of the fermions and the adjoint representation of

[48] While these operators form a Jordan algebra they do not have group representations.

he vector bosons makes the fermionic transformations such as eq. 3.1 'sluff' over internal symmetries.

3.3 Relation of the Θ-Symmetry Fields and the Fermionic Fields

The fermionic field operators that we have defined can be related to the U(192) Θ-Symmetry field operators which we will denote $A_{\Theta}{}^{\mu}{}_{nm}(x)$ where n = 1, ... , 192 and m = 1, ... , 192 are labels on the <u>192</u> fermion or boson fields. The following relations hold:

1. The product of $A_{\Theta F}{}^{\mu}{}_{nm}(x)$ and $A_{\Theta F}{}^{\mu}{}_{mp}(x)^{\dagger}$ can be expressed as an equivalent sum of Θ-Symmetry field operators.

2. The product of $A_{\Theta B}{}^{\mu}{}_{nm}(x)^{\dagger}$ and $A_{\Theta B}{}^{\mu}{}_{mp}(x)$ can be expressed as an equivalent sum of Θ-Symmetry field operators.

3. The product of $A_{\Theta F}{}^{\mu}{}_{nm}(x)$ and $A_{\Theta B}{}^{\mu}{}_{mp}(x)$ can be expressed as an equivalent sum of Θ-Symmetry field operators.

Thus we find that the U(192) Θ-Symmetry field operators can be obtained from products of the fermionic operators.

We also note that the product of a fermionic operator and a Θ-Symmetry field operator is a fermionic operator.

These relations establish a close connection between fermionic field operators and the Θ-Symmetry field operators.

We conclude that there is a SuperSymmetry relation between the fermion and vector boson multiplets in the Unified SuperStandard Model. The benefits of SuperSymmetric transformations in this context remain to be determined. However there is a hint of a significant role in the Megaverse. See chapter 5.

4. Shortcomings of Conventioal SuperSymmetry and SuperStrings

This chapter discusses the conventional approaches to SuperSymmetry and SuperStrings and their shortcomings.

4.1 Conventional SuperSymmetry

The original motivation for SuperSymmetry was the desire to have both fundamental fermions and bosons in the same multiplet. [49] A supersymmetry operator(s) was postulated that transformed boson states into fermion states. Local SuperSymetry was shown to have a number of advantages including the automatic adjustment of mass scales, finiteness of perturbation theory calculations to all orders, and a direct road to gravity.

However the formulations of Supersymmetric field theories ran into the simple problem that fermions are in fundamental representations while gauge fields are in adjoint representations. Consequently one could not simply combine the known fermions and gauge bosons into one multiplet. Instead physicists were led to combine the known fermions with new bosonic particles: squarks and sleptons, and the known bosons with new fermionic particles: Gauginos, Higgsinos, and goldstinos.

The original goal of SuperSymmetry was not met: the known fermions and bosons were not combined into a multiplet. And the many experimental searches for the new proposed particles have not been successful. It is possible that the proposed new particles may have masses much beyond the range of current accelerators. However the hope for this prospect is diminishing in view of the continuing disappointment as accelerator energies grow ever large without finding support for the proposed SuperSymmetry particles.

In contrast our Unified SuperStandard Model has equal numbers of vector bosons and fermions, which can be viewed as existing in two U(192) representation which can be combined in one $\underline{192} \oplus \underline{192}$ representation. Thus it embodies SuperSymmetry in a form similar to that initially expected!

4.2 SuperStrings

SuperString theories have attracted much attention because of the beautiful mathematics that they embody. However the concept of fundamental stirngs is obviously *much less*

[49] Myazawa, H., Prog. Theor. Phys. **36**, 1266 (1966); Phys. Rev. **170**, 1586 (1968); Gol'fand, Yu. A. and Likhtman, E. P., Sov. Phys.: JETP Lett. **13**, 323 (1971)

fundamental than the concept of a fuzzy space-time consisting of points made fuzzy by quantum effects: 2-dimensional constructs vs. fundamental 1-dimensional constructs. We point out later fuzzy points, as embodied in our quantum coordinates formulation, lead to finite results to all orders in perturbation theory and can be directly connected to the Standard Model as a quantum field theoretic extension.

SuperString theories, on the other hand, can only partially connect to the Standard Model through a chain of assumptions and approximations. Consequently, one must view SuperStrig theory either as a beautiful part of Mathematical Physics or as a theory that possibly may be relevant much into the future. With rare exceptions proposed 'future' theories have not been successful. Nature often turns out to be more creative – for example, quantum theory.

For the present it appears that the theory presented in volume I, and extended in this book, is a theory for the present and for the foreseeable future. What is known conforms to the theory. The theory makes numerous predictions as to the form of the fermionic and vector bosonic structure that should be susceptible to experimental tests in the foreseeable future.

5. Unified SuperStandard Model Extended to the Megaverse

5.1 Megaverse Dimension

In Blaha (2017c)[50] we developed a theory of the Megaverse – a higher dimensional complex space that contained universes such as ours. We set the dimension of the universe to 64 based on an argument that each of the 64 fields in the set of fields

$$E = SU(3) \otimes SU(2) \otimes U(1) \otimes SU(2) \otimes U(1) \otimes U(4) \otimes U(4) \otimes U(4)$$

corresponded to a Megaverse dimension. The argument was

> The motivation … can be discerned from considering a 2-dimensional space, and introducing a simple 1/r potential such as:
>
> $$V = g^2/(x^2 + y^2)^{-\frac{1}{2}}$$
>
> where g is a coupling constant.
> One views V as the potential of a force. However the values of V suitably extended to the range $[-\infty, +\infty]$ can be viewed as a third dimension.

We then proceeded to determine the features of the Megaverse using 'D' to denote the dimension of the Megaverse. We now use the number of fields in the Unified SuperStandard Model symmetry group – 192 fields:

$$[SU(3) \otimes SU(2) \otimes U(1) \otimes SU(2) \otimes U(1) \otimes U(4) \otimes U(4)]^4$$

to specify a Megaverse dimension

$$D = 192 \tag{5.1}$$

Almost all of the discussion of Blaha (2017c) remains true for this value of D. We will see that this choice adds a new 'dimension' to our discussion of Θ-Symmetry and SuperSymmetry.

[50] *Megaverse: The Universe of Universes.*

5.2 Extending Θ-Symmetry to Coordinates

The new choice of dimension, 192, enables us to extend the Θ-Symmetry toMegaverse coordinates. We define

$$y' = C_\Theta(y)y \qquad (5.2)$$

where y and y′ are 192-vectors and $S_\Theta(y)$ is a local Θ-Symmetry transformation. Then applying a Θ-Symmetry transformation to the Dirac equation (extended to Megaverse space)

$$\gamma_\mu D^\mu \psi = \gamma_\mu \{\partial^\mu + i\,[g_\Theta A_\Theta^{1\mu}(y) + g_\Theta A_\Theta^{2\mu}(y) + A_I^{1\mu}(y) + A_I^{2\mu}(y)]\}\psi(y) = 0 \qquad (2.9)$$

we find the Dirac equation eq. 2.9 transforms to

$$\gamma_\mu \{\partial^\mu + i\,[g_\Theta A'_\Theta{}^{1\mu}(y) + g_\Theta A'_\Theta{}^{2\mu}(y) + A'_I{}^{1\mu}(y) + A'_I{}^{2\mu}(y)]\}\,\psi'(y') = 0 \qquad (5.3)$$

where

$$A'_\Theta{}^{1\mu}(y') = C_\Theta(y)A_\Theta^{1\mu}(y)C_\Theta^{-1}(y) - i\,C_\Theta(y)\partial^\mu C_\Theta^{-1}(y)/g_\Theta \qquad (5.4)$$
$$A'_\Theta{}^{2\mu}(y') = C_\Theta(y)A_\Theta^{2\mu}(y)C_\Theta^{-1}(y)$$
$$A'_I{}^{1\mu}(y') = C_\Theta(y)A_I^{1\mu}(y)\,C_\Theta^{-1}(y)$$
$$A'_I{}^{2\mu}(y') = C_\Theta(y)A_I^{2\mu}(y)\,C_\Theta^{-1}(y)$$
$$\psi'(y') = C_\Theta(y)\psi(y)$$

Thus the Θ-Symmetry group becomes a coordinate and field transformation group. Since the the Θ-Symmetry group is U(192) it also plays the role of the Reality group for the Megaverse since it can transform any complex-valued coordinate system to a real-valued coordinate system.

The Θ-Symmetry interaction is experienced by all particles (fields) in the Megaverse. It is also experienced by all universes in the Megaverse and thus affects universe dynamics.

We choose the real-valued y^{192} coordinate to be the time coordinate. Then the Megaverse has a Complex Lorentz group with one time coordinate and 191 spatial coordinates.

Dynamics is invariant under the Complex Lorentz group transformations. Blaha (2017c) describes the possible motions of universes, and universe interactions, in detail.

Lastly, fermionic operators such as those considered in chapter 3 can be defined with similar features. Due to differences in internal symmetry representations, fermions and bosons cannot be united in a supermultiplet in the Megaverse.

5.3 SuperSymmetry and the Megaverse

The Megaverse offers an exciting new platform for SuperSymmetry studies of the Unified SuperStandard Model and possible SuperSymmetric generealizations using Megaverse Superspace with antisymmetric coordinates in addition to normal coordinates.

We can define a generalized set of Megaverse coordinates that associate 192 antisymmetric coordinates with the 192 normal Megaverse coordinates:

$$\{y^\mu, \Theta^\nu\}$$

where $\mu = 1, \ldots , 192$ *and* also $\nu = 1, \ldots , 192$. We now have the happy coincidence of equal numbers of normal and antisymmetric coordinates in Megaverse superspace.

Supersymmetric transformations on superspace coordinates have the infinitesimal form:

$$y^\mu \rightarrow y^\mu + i\,\overline{\varepsilon}\,\Gamma^\mu\Theta$$
$$\Theta^\nu \rightarrow \Theta^\nu + \varepsilon^\nu$$

(5.5)

where Γ^μ is a $192{\times}192$ matrix generalization of the Dirac γ matrices.[51] There are 192 Γ matrices.

Using Megaverse superspace coordinates one can proceed to construct a superfields formalism with many features similar to the 4-dimensional formulation.

[51] See Blaha (2017c) for their definition for arbitrary dimension D.

6. The New Quantum Field Theory

Quantum Field Theory as it has developed over the years has been constrained in a number of major ways that led to infinities in perturbation theory calculations and other puzzles. We have developed a generalization of Quantum Field Theory using a number of extensions that have resolved all the signifiacant issues that have led physicists to consider alternative approaches such as SuperString theories, which one must admit have problems of their own making. In this chapter we discuss the innovations that we have introduced over the past seventeen years.[52]

6.1 Complex Space-Time

Complex space-time has been a speculative subject, off and on, for many years. Its most signal success in the past was the proof of its necessity for proofs in axiomatic quantum flield theory. (See Streater (2000) for a detailed discussion.)

In volume I we showed that complex space-time led directly to the known periodic table of fermions consisting of four fermion species (expanded to 12 species taking account of 3 quarks in a quark species, and assuming 4 Dark fermion species). Through the Reality group the known Standard Model interactions emerged as the mechanism for mapping complex-valued coordinate systems to real-valued coordinate sysytems. Thus the Complex Lorentz group led directly to the form of the Standard Model, which had for many years been called 'bizarre.' A further reasonable extension of the model to the Generation and Layer groups led to the fermion and vector boson structure displayed on the book's cover.

An important consequence of the Complex Lorentz group is that it embodies faster-than-light speeds. Thus tachyons are implied. This feature opens the door to rapid starship travel as we ponted out in volume I and several earlier books.

Complex Spevial Relativity naturally generalizes to Complex General Relativiy which opens the door further to new interactions and new General Relativistic solutions. It also makes the concept of a Megaverse – a Universe of Universes – more palatable as discussed previously.

6.2 Two-Tier Quantum Field Theory

In Blaha (2002) we introduced a formalism for quantum coordinates that eliminated all infinities appearing in any order of perturbation theory calculations including the notorious fermion triangle diagram. It was based on making space-time points 'fuzzy' through a simple generalization of coordinates

[52] See voume I, Blsha (2002) and (2005a).

$$X_\mu(y) = y_\mu + i\, Y_\mu(y)/M_c^{\,2} \qquad\qquad (6.1)$$

where y_μ is a space-time coordinate, $Y_\mu(y)$ is a free field of the form of the electromagnetic field, and $/M_c$ is a large mass – perhaps of the order of the Planck mass that sets the scale of quantum smearing.[53]

Besides eliminating divergences in perturbative calculations it enables the possibility of higher dimensional quasntum field theories without divergences. Thus we can construct quantum field theories for the Megaverse.

Also quantum field theories, that are normally highly divergent, can also be constructed that are without divergences.

Two-Tier Quantum Field Theory thus opens the door to a vastly larger set of acceptable quantum field theories.

A final, and most important feature of Two-Tier theories, is that they preserve the successes of conventional quantum field theory. In particular, the astounding successes of Quantum Electrodynamics, which have calculated electrodynamic parameters with a precision beyond that of any other physical calculation, are maintained. (See the work of T. Kinoshita and colleagues.) Thus at energies much below M_c existing quantum field theory results remain true in Two-Tier quantum field theory.

6.3 Cordinate System Covariant Particle Interpretations of States

Unruh[54] and many others have pointed out that conventional second quantization in accelerating, and other exotic coordinate systems, results in ambiguous asymptotic particle states interpretations. In one coordinate system an n particle state appears as a superposition of multiparticle states in another coordinate system. Some years ago this author[55] pointed out that a generalization of quantum field theory which admits Bogoliubov transformations of creation and annihilation operators enables the definition of states whose particle interpretation is unambiguous under an arbitrary change of coordinate system. An n particle state in one coordinate system remains an n particle state in a different coordinate system.

Thus the use of our prposed generalization of quantum field theory results in a physically acceptable quantum field theory with unambiguously defined asymptotic states.[56]

[53] See Blaha (2005a) for a detailed discussion.

[54] Unruh, W. G., Phys. Rev. D **40**, 1053 (1989) and references therein.

[55] S. Blaha, "The Local Definition of Asymptotic Particle States", IL Nuovo Cimento **49A**, 35 (1979); S. Blaha, "New Framework for Gauge Field Theories", IL Nuovo Cimento **49A**, 113 (1979).

[56] A reader might suppose that this issue is a 'paper tiger' with no practical relevance. However there has recently been a continuing discussion on the topic of whether a rapidly accelerating space ship might encounter clouds of particles and a 'heated' vacuum that would damage it. Our new theory shows that this possibility will not happen. Space exploration will not be hampered by this pseudo-problem.

Interestingly the formalism that we develop for this potential situation enables us to *canonically* define higher derivative quantum field theories (See section 6.4.) for quark confinement and other purposes such as modifications of gravitation from a 1/r potential (MOND-like). It also enables us to define a new Higgs sector quantum field theory with a clean separation between Higgs vacuum expectation values and their associated quantum field parts. (See section 6.5.)

6.4 Higher Derivative Quantum Field Theories

Using a formalism that associates two quantum fields with each vector boson, PseudoQuantum Field Theory, we showed in volume I and in earlier books that one could create a higher derivative quantum field theory from a lagrangian using the standard canonical quantum field procedure.

We applied this formalism for the case of the Strong Interaction to derive a linear r potential of the form that was successfully used in the calculation of the charmonium spectrum.

We also applied the formalism to gravitation and showed that the gravitational 1/r potential was replaced by three gravitation potentials: a 1/r potential at solar system scales, a modified potential at intra-galactic distances, and yet another potential at inter-galactic distances. These potentials were in qualitative agreement with data obtained in experimental analyses. Thus in both the Strong Interaction and gravitation cases we found a higher derivative theory derived through canonical methods agreed with experiments.[57]

6.5 Reformulation of Higgs Boson Theories

The two field per particle formalism described in sections 6.3 and 6.4 above also provides a formalism for the Higgs Mechanism, which the author believes is cleaner and better than the usual formalism. The basic idea is to define one field φ_1 as the field that produces the symmetry breaking vacuum expectation value[58]

$$\varphi_1(x) = \varphi_{10} + \varphi_1'(x) \qquad (6.2)$$

where φ_{10} becomes the vacuum expectation value when $\varphi_1(x)$ is applied to a coherent[59] vacuum state. $\varphi_1'(x)$ is a 'normal' quantum field. The second field $\varphi_2(x)$ is a quantum field that has a non-zero commutator with the canonical momentum of $\varphi_1(x)$.

Thus we are able to cleanly separate the vacuum expectation value from the quantum fields associated with a Higgs particle. This is described in detail in volume I.

[57] See volume I for details and references.

[58] As shown in volume I and earlier books $\varphi_1(x)$ commutes with its canonically conjugate momentum.

[59] Coherent states are well known in the physics literature. See for example T. W. B. Kibble, J. Math. Phys. **9**, 315 (1968) and references therein; V. Chung, Phys. Rev. **140**, B1110 (1965); J. R. Klauder, J. McKenna, and E. J. Woods, J. Math. Phys. **7**, 822 (1966) and references therein.

6.6 New Quantum Field Theory

The new formulation that we have proposed resolves the problems hitherto associated with quantum field theories, and obviates the need for the consideration of alternatives for the foreseeable future.

7. Construction Principles for a Physical Theory of Universes

Euclid was fortunate in the creation of Euclidean geometry because he knew the method with which to construct geometry, and the primitives, straight lines, curves, and angles in space with which to define his axioms. The development of a fundamental physical theory is not so fortunate. The fundamental principles and axioms are open to debate. Many proposals have been put forward. They can be classified into two general categories: fundamental theories which simply assume a set of fundamental particles and symmetries; or fundamental theories which assume a set of fundamental artifacts such as strings and symmetries that ultimately form the known elementary particles and symmetries of the Standard Model. Both approaches are open to the same questions: What is their justification? What is their origin? How do we know that there is not yet a more fundamental level?

We think that we have derived a fundamental theory of the most primitive possible type that directly leads to the Standard Model with a natural extension for Dark Matter, and Quantum Gravity. Its origin is Special and General Relativity. From this basis we developed the spectrum of fundamental fermions and vector bosons with the Higgs Mechanism used to generate symmetry breaking and particle masses.

In this chapter[60] we will consider certain fundamental general concepts which set directions at points in the construction where there was a choice.

7.1 A Principle of Logic as the Origin and the Method of Construction

7.1.1 Requirement of Logic in Physics

Every physical theory is developed by logical deduction. Questions have been raised about the validity of logic. In particular, Gödel's Undecidability Theorem[61] is often cited as proof of a fundamental problem in logic. As we pointed out in Blaha (2011c), "We show these "paradoxes" and the Undecidability Theorem are resolved if we understand logic statements to be a form of function whose function arguments have a domain of validity. The verbal and symbolic statements, that are at the heart of paradoxes, use 'function' arguments that are outside

[60] This chapter, and chapters 8 and 9, largely appeared in Blaha (2015a).
[61] Gödel's Undecidability Theorem is not incorrect. However its interpretation, as a flaw in the nature of Logic, is incorrect. It really is a proof of the existence of statements that are both true and false because of the choice of subjects outside the domain of predicates. We view statements as a 'verbal' form of function.

the domain of the functions embodied in their statements.[62] Invalid arguments are the source of these seeming paradoxes and Gödel's Undecidability Theorem." Thus Logic is valid and not internally inconsistent.

We therefore adopt the principle that a fundamental theory must be logical.

7.1.2 The Logic Building Block of Particles – The Iota

Furthermore if we consider all possible 'things' that might constitute a fundamental building block for a fundamental theory they are all, at best, *ad hoc*, and raise questions of their necessity, and also whether they are composed of yet a more fundamental substructure.

There is only one choice of building block that avoids these issues – a logic unit or bit. A unit of logic is a fundamental entity that is known to have an energy or equivalently a mass, and has no constituents of a more primitive form.[63] We will call a unit of logic that will form the core of a particle an *iota*.It has a conceptual value. But, in itself, it has no physical form or material existence in the sense of lumps of matter unless features, such as coordinates, and supporting interactions are introduced by construction. Later we will introduce physical features that will cloak iotas with properties and interactions.

7.1.3 Mass of an Iota

Recent experiments have shown that a logical value (a bit or iota) has an energy.associated with it. One bit of information has about 3×10^{-21} joules of energy[64] or a rest mass, m_0, or about 0.02 eV using $m_0 = E/c^2$. This result was confirmed by Eric Lutz et al.[65] who showed that there is a minimum amount of heat produced per bit of erased data. This minimal

[62] Most scientists, Logicians and Mathematicians are familiar with mathematical functions that have numerical arguments and calculate a numerical value or set of numerical values. However, those familiar with Computer Science know that a function, in general, has two types of output: its numeric value(s), and a status value indicating whether the computation that the function performed was successfully done or failed. Typically the status value is a zero or one, but it is understood as true or false also. (True – computation successful; false – computation failed for some reason) Thus we categorize functions as being in one of three broad categories:

Types of Functions

1	2	3
A Mathematical Function Producing A Value(s) Only;	A Function Producing a Value(s) and a Status Value;	A Function Producing a Status Value Only;
Examples:		
sin function	Programming Function	Logic Statement

[63] An iota is a 'physical' manifestation of a logical value. The relation of an iota to a logical value is analogous to the relation of a penciled point placed on paper to the concept of a point as a primitive in geometry.

[64] E. Muneyuki et al, *Nature Physics*, DOI: 10.1038/NPHYS1821.

[65] E. Lutz et al, Nature **483** (7388): 187–190,10.1038/nature10872, (2012).

heat is called the *Landauer*[66] *limit*. The equivalent mass we will call the *Landauer mass*, and we will denote it as m_0. We assume that a fundamental Landauer mass exists in our discussions although the precise value of the mass will not be used since we may expect all physical particle masses to be renormalized to different values when interactions are taken into account.

However, it is intriguing that the mass of the electron neutrino has been measured in a variety of experiments and found to be within an order of magnitude or so larger than the Landauer mass estimate. We expect this since particles acquire a 'cloud of virtual particles' due to interactions that increase their mass above the Landauer mass. Since neutrinos only have the weak interaction it is not surprising that the increase due to interactions should not be large. The Mainz Neutrino Mass Experiment, for example, estimates the electron neutrino mass to be less than 2 eV.

A number of astronomical studies have also generated estimates of neutrino masses. In July 2010 the 3-D MegaZ DR7 galaxy survey found a limit for the combined mass of the three neutrino varieties to be less than 0.28 eV.[67] A smaller upper bound for the sum of neutrino masses, 0.23 eV, was found in March 2013 by the Planck collaboration,[68] In February 2014 a new estimate of the sum was found to be 0.320 ± 0.081 eV due to discrepancies between the Planck's measurements of the Cosmic Microwave Background, and other predictions, combined with the assumption that neutrinos are the cause of weaker gravitational lensing than implied by massless neutrinos.[69]

Thus the experimentally measured values of neutrino masses are consistent with the iota Landauer mass estimate of 0.02 eV given above. We can thus assume that a particle consists of an iota with a certain mass[70] *that is renormalized, and having other features, that will emerge in the construction of the complete theory.*[71] We view reality as ultimately a representation (or painting) of logic values evolving through interactions in time and space.[72]

[66] R. Landauer, "Irreversibility and heat generation in the computing process", IBM Journal of Research and Developm
ent **5** (3): 183–191, (1961).

[67] S. Thomas et al, "Upper Bound of 0.28 eV on Neutrino Masses from the Largest Photometric Redshift Survey", Physical Review Letters **105**: 031301 (2010).

[68] Planck Collaboration, arXiv:1303.5076 (2013).

[69] R. A. Battye et al, "Evidence for Massive Neutrinos from Cosmic Microwave Background and Lensing Observations", Phys. Rev. Lett. **112**, 051303 (2014).

[70] Leibniz first proposed the idea of logic 'particles' which he called monads. Our definition of a logic 'particle' does not include (or exclude) the presence of a spiritual part which was part of the definition of Leibniz's monads.

[71] A recent experiment claims to separate the spin part (which we identify as a logical value later) of a molecule from the rest of the molecule.

[72] Those who might suggest matter is substantial, and logic values are not, should remember that matter would be completely insubstantial if there were no forces in nature. Neutrinos which are close to insubstantial would be completely insubstantial if there were no weak interactions.

7.2 Ockham's Razor

In the construction of a theory there are occasions when a choice must be made between several alternatives. William of Ockham proposed a Law of Parsimony that is called *Ockham's Razor* which states that the simplest choice is to be preferred in a multiple choice situation. This principle is often stated as 'the simplest solution is usually the correct solution.'[73]

The best rationale for this principle is that such a choice generally reduces the complexity that amost always follows when computations are performed. Since many physics calculations are extremely difficult, picking the simplest choice.would generally tend to make subsequent theorems/calculations less difficult. This point of view might be thought to be ad hoc or anthropomorphic. But it reflects the reality of scientific calculation and of theory construction.

Thus we will assume Ockham's Law of Parsimony in the construction of our theory with the proviso that Leibniz's Minimax Principle (section 7.3) takes precedence if there is a conflict in the implications of the choices.

7.3 Leibniz's Minimax Principle for Physics Theories

Leibniz[74] developed a Minimax Principle that can be phrased for our purposes as, "The universe is based on the smallest set of properties or features that lead to the greatest variety of phenomena." This principle reflects the spirit of the minimum/maximum criteria of the Calculus of Variations[75] that play a central role in many physics theories. This principle somewhat overlaps Ockham's Law of Parsimony. Given a choice of possible theoretical lines of construction there is a possibility that the Law of Parsimony and the Minimax Principle would suggest different choices. Fortunately, we will see that the construction of the Extended Standard Model does not seem to present this potential dilemma.

An important, unremarked, aspect of Leibniz's Principle is the decision between a set of choices depends on the 'future' of the construction or theory. Thus future constructs determine past constructs in minimax decisions.[76]

7.4 A Physical Space Exists

We will postulate that a space exists of as yet undetermined dimensions and parametrized by coordinates. We will assume that an invariant distance measure can be defined

[73] William of Ockham – Law of Parsimony – "Pluralitas non est ponenda sine necessitate" or "Plurality should not be posited without necessity." Ockham's Law was first stated by Durand De Saint-Pourçain (1270-1334 A.D.). In simple terms the principle states the simplest solution to a problem is most likely to be the correct solution.

[74] See Rescher (1967).

[75] Leibniz was one of the founders of the Calculus of Variations.

[76] The knowledgable reader will remember Feynman's speculation that the physical universe may be evolving from the future into the past. Quantum field theories support such an interpretation of their mathematics. The similarity of this fundamental minimax principle's feature with the corresponding feature of physics theories encourages support for the minimax principle as a 'design' law of fundamental physical theory.

on that space. Physically measured coordinates are necessarily real-valued because clock time, and ruler, measurements always yield real values.

7.5 Non-Localized Physical Processes Exist

Physical processes can exist with spatially separated parts that may be intimately coordinated. Processes with spatially separated parts can have instant communication between the parts or can have parts communicate with time delays. (Quantum Entanglement)

7.6 Construction Principles of Euclid's Geometry

In the above sections we have discussed construction principles for a fundamental physics theory. One might ask why does not Euclid's geometry have analogous construction principles? Normally we think of Euclid's geometry as based on primitive terms and five postulates.[77]

With small thought we see that Euclid's geometry has construction principles as well *at an intuitive level.* These principles include the definition of angles, the construction of simple geometric figures such as rectangles, triangles, trapezoids, and so on. The more advanced developments in geometry such as the geometry of Ptolomaic astronomy with cycles and epicycles also have implicit principles for construction.

Thus our introduction of additional principles 'along the way' to guide the construction of fundamental physics is analogous to the implicit construction/derivation principles of Euclid's Geometry.

[77] Construction principles such as building triangles and other geometric figures as well as dissecting their features are implicit. Also there are some hidden assumptions, beyond the original five, that subtly appear in the diagrams used in the proof of theorems.

8. Asynchronous Logic and Physical Processes

In sections 7.1 and 7.5 of chapter 7 we discussed the central role of Logic and the need for synchronization of non-local physical processes. The need for synchronized non-local physical processes requires the introduction of a new principle: the Principle of Asynchronicity.[78] When processes take place in parallel whether it is Quantum Mechanical entangled processes at small/large distances, or in high order Feynman diagrams (or their old fashioned time ordered perturbation theory predecessor) the synchronicity of a process is a physical requirement. It is implicitly resolved by physical laws which prevent asynchronicities (situations when parallel processes get "out of sync" resulting in the failure of an entire physical process to complete properly.) The Principle of Asynchronicity is described in the following pages. Asynchronicity can be briefly described as:

In computation asynchronicity issues can arise. For example parallel computations or computer processes on a chip or set of chips have to be carefully managed for a parallel computer process to complete properly. In the case of computer chip design (VLSI chips and so on) techniques have been developed for the design of chips based on multi-valued logic. One conceptual approach uses 4-valued logic to define clockless computer logic circuits. The 4-valued logic developed by Fant (2005) has the four logic values TRUE, FALSE, NULL, and INTERMEDIATE. It is an extension of Boolean Logic that can accommodate time asynchronicities in asynchronous computer circuits. It enables circuits to avoid the use of system clocks to implement synchronization.[79] Thus the synchronization is explicit in 4-valued logic and non-logical constructs are not needed.[80] Concurrent transitions are coordinated solely by logical relationships with no need for any time constraints or relationships.

Now, realizing that The Standard Model, and physical theories that are ultimately derived from it such as Quantum Mechanics, potentially contain asynchronicities, we suggest that a Principle of Asynchronicity is embodied in the fundamental theory of Physics that leads

[78] Much of this chapter is abstracted from Blaha (2011c) and printed in smaller type. Some might argue that it should be called the principle of synchronicity since the goal is synchronization of the parts of an evolving process. We chose to follow the terminology in the field of Asynchronous Logic as exemplified by Fant (2005) – a classic in that field.

[79] Remarkably Bjorken (1965) pp 220-226 presents an analogy of Feynman diagrams with electrical circuits where momenta map to currents, coordinates to voltages, Feynman parameters to resistance, and free particle equations of motion to Ohm's Law plus the equivalent of Kirchhoff's Laws. Thus Feynman diagrams and computer circuits are completely analogous.

[80] A two-valued asynchronous logic is also possible – just as the Dirac equation can be expressed as two 2-dimensional equarions. See Fant (2005) and Bjorken (1965).

:o Dirac-like equations for the fundamental fermions – the leptons and quarks of The Standard Model.

8.1 Four-Valued Asynchronous Logic

The basic defining features of asynchronous circuits and Asynchronous Logic are:

1. An *asynchronous circuit* is a circuit in which the component parts are autonomous and can act in parallel at various rates of time evolution. They are not controled by a clock mechanism but proceed or wait for signals indicating that they can proceed.
2. *Asynchronous logic* is the logic used in the design of asynchronous circuits. The logic embodies the asynchronicity, and so the circuits built using the logic do not use a clock to control the execution speed of the various parts of an asynchronous circuit. Consequently logic elements do not necessarily have a distinct true or false state at any given point in time. 2-valued Boolean Logic is not sufficient and so asynchronous logic is multi-valued. The logic embodies states that allow for "stop and go" states within an executing asynchronous circuit.

In Fant's asynchronous 4-valued logic the four possible truth values of a state are:

True – status is true and all data is current
False – status is false and all data is current
Intermediate – status is indefinite with some data current
NULL – status is indefinite with no data present – results in a suspension of processing of the circuit part in a NULL state until current data becomes present

"Data" is the information flowing through all or part of a circuit. Using these truth values the evolution in time of the parts of an asynchronous circuit are effectively synchronized by the logic without the use of a clock mechanism. (A clock mechanism effectively is a subsidiary time constraint or set of time constraints.) See Fant (2005) for further details.

An implicit aspect of asynchronous logic is the coordination of spatially separated parts of a circuit. Since spatial separations in a circuit can be mapped to time delays using the speed of data propagation between parts, spatial asynchronicites are subsumed under time asynchronicities. This is particularly true for computer chips which are kept small to minimize delays.

8.2 Principle of Asynchronicity

An obvious feature of elementary particle phenomena is the coordination of the parts of a physical process in time and space. Complex Feynman diagrams embody the coordination of the parts of interacting particles. Quantum entanglement phenomena embody the coordination of the parts of a physical phenomena separated by large distances and perhaps times. Examples of these types, which could be multiplied indefinitely, lead to a Principle of Asynchronicity.

Principle: Nature requires asynchronicity. This asynchronicity is coordinated by 4-valued physico-logical structures for matter.

Elaboration: Elementary particle physical phenomena must support extended coordinated physical phenomena in space and time. The fundamental laws of particle physics must be such

as to permit coordinated physical phenomena with coordination between the parts of a physical phenomenon at small/large distances and small/large time intervals. The coordination must be embodied within physical laws.

This principle will be applied below (and later in this book in more detail) to justify Dirac-like equations for particle dynamics.

Coordination is an obvious feature of physical phenomena. This principle goes beyond that by asserting that extended coordinated physical phenomena must exist. If particles exist, then their antiparticles must also exist to provide asynchronous behavior in interaction regions. If only particles existed then all interactions would proceed forward in time and the state of the interaction at any point in time would be known. With the addition of antiparticles, asynchronicity issues are introduced and at various 'time slices' (if one thinks in terms of old-fashioned time ordered perturbation theory) of the progress of an interaction, the state can be ambiguous since antiparticles are negative energy particles moving backward in time.

Asynchronicities are common in the many subcircuits of a computer chip. Asynchronicities are also common in the many interaction subregions of a set of particles in interaction. Page 7 of Fant (2005) has a diagram of a circuit with a set of subcircuits with five time slices of the interacting subcircuits showing five states of the "'data' wavefront" at five points in time. This diagram is similar to the time-sliced diagram of an interacting system of particles in "old fashioned" time-ordered perturbation theory. Page 29 of Blaha (2005b) displays a similar diagram (Fig. 5.1.4) in a description of a Standard Model Quantum Langauge Grammar – a language representation of particle physics. Blaha's diagram[81] is remarkably similar to Fant's diagram in overall features as one might expect since both address time asynchronicity.

The asynchronicity that appears in perturbation theory diagrams is intimately related to the appearance of antiparticles in diagrams. As noted earlier antiparticles are interpretable as negative energy particles traveling backwards in time. The time orderings, which are implicit in the Feynman diagram approach and explicit in old fashioned perturbation theory, evidence time asynchronicity and the effects of the dynamics. They coordinate asynchronicities such that correct results follow from perturbative calculations.

[81] Created without knowledge of Fant's work.

9. Iotas as the Core of Fermions

9.1 Matrix Representation of Asynchronous Logic

The four possible logic states of Asynchronous Logic can be mapped to a matrix representation with four component columns and 4×4 matrices that transform between logic values.[82] The basic four pure logic states can be labeled using a notation that connects with physics:

$$u(+\tfrac{1}{2}) = \begin{bmatrix} 1 \\ 0 \\ 0 \\ 0 \end{bmatrix} \qquad u(-\tfrac{1}{2}) = \begin{bmatrix} 0 \\ 1 \\ 0 \\ 0 \end{bmatrix} \tag{9.1}$$

and

$$v(-\tfrac{1}{2}) = \begin{bmatrix} 0 \\ 0 \\ 1 \\ 0 \end{bmatrix} \qquad v(+\tfrac{1}{2}) = \begin{bmatrix} 0 \\ 0 \\ 0 \\ 1 \end{bmatrix} \tag{9.2}$$

The arguments of u and v will become physically values of particle spin: $+\tfrac{1}{2}$ represents an "up" spin state and $-\tfrac{1}{2}$ represents a "down" spin state. Linear combinations of these four states obtained using linear combinations of the sixteen 4×4 matrices with complex coefficients form *qubits*.[83] Any bit or qubit transformation can be constructed from a sum of the sixteen Dirac matrices[84] multiplied by complex coefficients. The iota states listed above are constants. They will become variable through the use of Lorentz transformations considered later.

A set of sixteen independent 4×4 matrices can be constructed from the four basic Dirac matrices, usually denoted γ^μ, by multiplications and summations. They can be found in most books on quantum field theory.

[82] See Blaha (2010a) for more details.
[83] S. Weisner, "Conjugate coding". Association for Computing Machinery, Special Interest Group in Algorithms and Computation Theory **15**, 78–88 (1983). We will not discuss qubits in this construction.
[84] See Bjorken (1964).

Applying these basic facts about the 4×4 matrix representation of Asynchronous Logic to iotas we can develop the Dirac equation formulation of spin ½ particles after introducing space-time coordinates. See volume I.

We can view fermion particles as iotas parametrized by coordinates and interacting via interactions (forces) that primarily result from the need to impose symmetry requirements on coordinate transformations.[85]

9.2 Matter is Insubstantial Neglecting Particle Interactions

Philosophers and physicists have debated the nature of matter for milleniums. Differing opinions have been the norm. Perhaps one of the most interesting expressions of opinion was that of Dr. Johnson, the 19[th] century author of the Dictionary. Upon hearing of Bishop Berkeley's philosophic view that matter was insubstantial and "not real", Dr. Johnson proceeded to kick a rock while exclaiming "I refute it thus" according to his biographer Boswell. While Dr. Johnson's riposte cannot be denied in its succinctness, our theoretical development is based on the concept of iotas as the core of fermion particles, and the further development of our construction later leads to a refinement of the Berkeley-Johnson conflict as well as that of many philosophers and physicists.

For we propose that matter is truly insubstantial with Bishop Berkeley, and yet gains substantiality through interactions (forces) without which all matter would be interpenetrable and could reside at a single point.[86] Thus Dr. Johnson does not truly refute Dishop Berkeley but does show that forces exist amongst matter that gives it "substantiality."

Since all matter might have been concentrated[87] at an 'essentially' mathematical point in the absence of interactions we can call that point the *void* and view it as the 'location' of the Big Bang that presumably existed at the Beginning. Coordinates and interactions may well have originated at that point as the result of quantum fluctuations.

[85] We note that Asynchronous Logic can also be formulated with two dimensional vectors and 2×2 matrices just as the Dirac matrix formalism for spin ½ particles can be expressed in a 2×2 matrix formalism.

[86] To some extent we see an approximation of this proposal in the approximate interpenetrability of normal matter and Dark Matter which would be exact if there were not a very weak force between matter and Dark Matter as well as the gravitational interaction.

[87] This comment is based on our version of quantum field thory in which all interactions (including gravity) go to zero at zero distance. See Blaha (2005a) for a detailed discussion of our divergence-free form of quantum field theories. With zero forces, all particles can concentrate at a single point.

Appendix 9-A. A Map between Interacting Particle Theories and Computer Languages

This appendix appears as Appendix B in Blaha (2013b). It is also presented in more detail in Blaha(2005b) *The Metatheory of Physics Theories, and the Theory of Everything as a Quantum Computer Language* and in Blaha (2005c) *The Equivalence of Elementary Particle Theories and Computer Languages: Quantum Computers, Turing Machines, Standard Model, Superstring Theory, and a Proof that Gödel's Theorem Implies Nature Must Be Quantum.*

This appendix can be skipped by the reader not interested in a discussion of a fairly detailed comparison of elementary particle theory "in the raw" and computation.

In this appendix we will provide detailed support for the view that The Standard Model is a particular implementation of a general specification for a computation. In this view particles are skeletonized to data, and interactions skeletonized to the execution of a computation. Thus we provide a complementary view to the current effort to build quantum computers from atoms and molecules using Quantum Theory. The discussion in this chapter is largely condensed extracts of parts of Blaha (2005b). Our earlier works, beginning with Blaha (1998), contain the basic ideas and their elaboration.

9-A.1 The Similarity of Data to Particles

Some years ago we pointed out that true creation and annihilation in our universe appears in only two arenas:[88] data transformations within computers and particle transformations in Nature. Both subatomic particles and data can be created or destroyed or combined to produce new particles or new data. The similarity is compelling! In the following sections we will amplify these ideas based on the theory of computer languages and computer grammars.

We will also consider particle interactions at a more abstract level and show how to create "sub-atomic particle quantum Turing machines" for both the Standard Model and Superstring theories. (In this appendix we take a neutral stance on whether the Theory of Everything is a Superstring theory or our Extended Standard Model. The possibility of a deep connection of these two theoretic approaches cannot be ruled out at present.)

[88] Blaha (1998). Please note that we will discuss only Nature in this chapter since theological constructs are beyond the scope of our inquiry.

9-A.2 Particle Physics Lagrangians Defines a Language

Physicists use perturbation theory to perform computations in quantum field theories such as The Standard Model. Perturbation theory takes the interaction terms of the Lagrangian and performs approximate calculations of the probabilities of particle interactions. Perturbation theory calculations are normally visualized using Feynman diagrams. For example the collision of two electrons to produce two electrons with different energies and momenta (called electron-electron elastic scattering) can be visualized as a sum of terms corresponding to the various ways the electrons can interact. The number of terms in this example is infinite in principle. Since this infinite sum cannot be calculated the sum is approximated by a finite number of terms. The simplest and, as it turns out, the dominant terms in electron-electron scattering are:

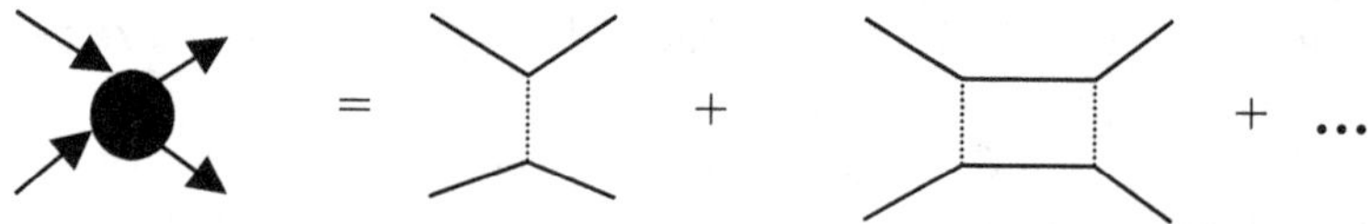

Figure 9-A.1. The first few terms in the approximate perturbation theory calculation of the scattering of two electrons. The dotted lines represent photons that carry the electromagnetic force between the electrons.

We will see that the individual terms (diagrams) in a perturbation theory calculation can be viewed as words. The words are part of a language with an alphabet and grammar.

The fundamental particles of the Standard Model constitute the set of symbols or the alphabet of a language. (It is an interesting, but perhaps a meaningless, coincidence that most alphabet-based human languages have 20 to 40 letters in their alphabets. English has 26 and so on.) As we shall see, the particle language grammar is a quantum extension of a type of computer language (developed by Chomsky and others) that uses *production rules*. The production rules for the grammar of the Standard Model are easily derived from the interaction terms of the Standard Model.

The concept of the language of the Standard Model is very simple. The technical details of the language description require a discussion of computer languages and Quantum Turing Machines.

The following sections describe the basic idea of this linguistic representation of the Standard Model. They show how particle interactions can be viewed as transformations (processes) within a Quantum Computer that accepts the computer language generated by the Standard Model Lagrangian interaction terms.

9-A.3 A Linguistic Representation of the Standard Model

In this book and earlier works we explored the features of the Extended Standard Model. We saw that it was consistent with almost all the known properties of elementary particles. This section introduces a new view or representation of Standard models that focuses on interactions and *shows that Standard Models define languages similar to computer languages.*

The *alphabet* of a Standard Model language is the set of elementary particles of the Standard Model. The *words* of the language are quantum states consisting of elementary particles. A bound state of several particles such as a proton (three quarks bound together) is a word. A quantum state consisting of several free particles – particles that are not bound together and that are some distance from each other also constitute a word. In fact the entire universe constitutes one mighty word.[89]

The collision or scattering of particles can be viewed as beginning with a combination of letters corresponding to the set of initial particles – the input string or word. This input string undergoes transformations specified by grammar rules to produce an output string (word) of letters corresponding to the outgoing particles after the collision.

In describing this new view of Standard Models we will focus on the essentials of the processes of creation, transformation and annihilation of matter ignoring (for the moment) particle spin, momentum, angular momentum and other details that are important in the complete theory of the Standard Model. This approximation may have been valid prior to the Big Bang when the universe might have been a mathematical point. Incorporating particle spin, momentum, angular momentum and so on into a "particle" language is not difficult.

The idea of associating physics with computers is not as unconventional as it might appear at first. Feynman[90] viewed computers as relevant for Physics: "If we suppose we know all the physical laws perfectly, of course we don't have to pay any attention to computers. It's interesting anyway to entertain oneself with the idea that we've got something to learn about physical laws; and if I take a relaxed view here … I'll admit that we don't understand everything." Feynman wanted to simulate physics computations on a quantum computer in the hope that it would be faster than a conventional computer. We will show the Standard Model itself actually defines a specific (theoretical) quantum computer – a far more exciting possibility – because it gives a new view of Reality. Nature itself is a form of computer. SuperString theory can also be formulated within a Quantum Computer framework.

A computer language representation of particle physics is of great interest in itself. It may generate new insights into the process of matter creation and transformation. It may lead to a new understanding of the fundamental nature of the universe. And it appears to suggest a rationale for approaches such as the currently popular theories of elementary particles.

[89] Blaha (1998).
[90] R. P. Feynman, International Journal of Theoretical Physics, **21**, 467 (1982).

9-A.3.1 Linguistic View of an Interaction

We will begin by looking at the simple interaction term:

$$\overline{e}Ae$$

From a computer language perspective this Lagrangian interaction term can be viewed as specifying a set of grammar rules called *production rules*.

In fact each interaction term in a Standard Model Lagrangian can be viewed as specifying a set of grammar rules. The combined set of grammar rules for all Standard Model interaction terms defines a grammar with particles constituting the alphabet (letters or symbols) of the grammar.

To appreciate the mapping (or analogy) between particles and alphabetic letters, and of interaction terms and computer grammar, we have to understand the process of data characters (or letters) flowing through a computer. It is an interesting and little noted fact (because it is viewed as trivial) that a computer can generate (or absorb) data as part of the computation process. For example we might write a computer program that takes a set of letters input into a computer and outputs each input letter twice:

In a sense the computer has created data characters just like particle interactions can create particles. Computers can also absorb (or annihilate) data (usually to our dismay). So we can see an analogy between the transformations of data characters in a computer, and particle annihilation and creation. *Nothing else in Nature is so directly analogous to particle creation and annihilation.*

This observation leads us to take the view that particles are data packets that we denote with letters (symbols). They contain quantum numbers and other properties (mass, spin, momentum, and so on) that certainly are data. And they have a grammar that we summarize with a Lagrangian.

9-A.3.2 Computer Grammars

A Standard Model Lagrangian in our view specifies a *grammar* in the sense of Naom Chomsky. Chomsky's concept of a language, and of a grammar, has important applications in the theory of computation and computers.

There are four basic types of languages in the Chomsky approach: called type 0, type 1, type 2 and type 3. They differ in the allowed forms of their grammar rules (also called *production rules*). We will be interested in type 0 languages. A type 0 language (also called an

unrestricted rewriting system) is the most general type of language. It allows any grammar production rule of the form

$$ x \longrightarrow y $$

where x and y are strings of characters.

Production rules specify how one string of characters transforms into another string of characters. Calculations in computers using computer languages are reducible to sets of grammar rules for string manipulation that are similar to the one shown above.

Each term in the interaction part of the Standard Model Lagrangian is equivalent to one or more production rules where the characters are particles. *A Standard Model can be viewed as generating a type 0 language.* This language goes beyond current types of grammars because it is inherently quantum probabilistic in nature. Quantum aspects of these rules will be described later.

Before looking at the production rules generated by an interaction term in a Lagrangian we will discuss a formal grammar. A grammar is a quadruple of items that is usually symbolized by the expression

$$ \langle N, \ T, \ S, \ P \rangle $$

where N is a set of variables called *nonterminal symbols*, T is a set of *terminal symbols*, S is a special nonterminal symbol called the *head* or *start symbol*, and P is a finite set of production rules. The angle braces < and > are merely a mathematician's way of saying these items are grouped together to constitute (or make) a grammar.

The *terminal* symbols are the set of characters that are allowed in input strings or output strings. The *nonterminal* symbols are the set of characters that appear in intermediate steps that lead from the input string to the output string. They are like internal variables or symbols.[91] The combined set of terminal and nonterminal symbols constitute up the *vocabulary* (alphabet) of a language.

Chomsky's definition of a language is the set of all strings of terminal symbols that can be generated by applying the production rules to the *head* symbol (or *start* symbol) S.[92] The head symbol is the symbol that begins all strings of symbols that can be generated in a language.

A simple example of a language in this approach is a vocabulary or alphabet consisting of the ABC's with words created from these letters according to some set of production rules.

[91] Their particle analogues are virtual particles that only appear in an evolving interaction but not in the input or output. One could view quark-partons as equivalent to nonterminal particles.

[92] The head symbol is analogous to the vacuum state in quantum field theory. Chomsky's definition specifies the possible vacuum fluctuations that can occur. Each string is a specific vacuum fluctuation. An example is an electron-positron pair momentarily popping out of the vacuum—a vacuum fluctuation.

9-A.3.3 Generalized Input Chomsky Languages

We will generalize Chomsky's idea of language to be the set of all strings that can be generated from all finite input strings of terminal symbols as well as the *head symbol*. We can also view all particles as generated directly or indirectly at the beginning of the universe. The "Big Bang" (the beginning of the universe) then becomes the primeval head symbol.

We can visualize the application of production rules to transform an input string of terminal characters into an output string of terminal characters as:

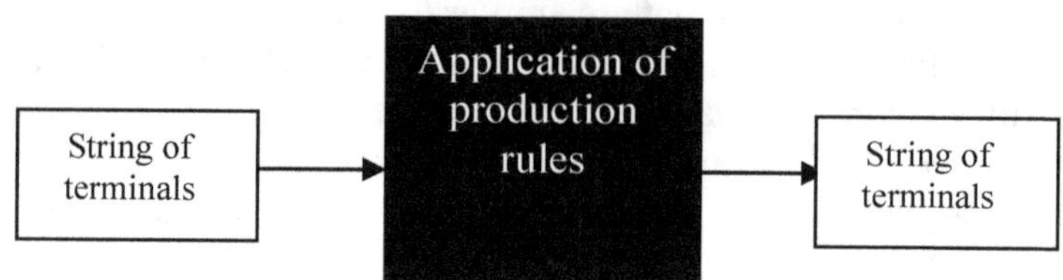

Figure 9-A.2 Generating an output string of terminal characters from an input string of terminal characters using the production rules of a grammar. Inside the black box, transformations of the input string take place and non-terminal symbols may appear and disappear. Non-terminal symbols, by definition, cannot appear in the input or output strings of characters.

In order to make these grammar concepts more concrete we will look at a simple artificial grammar before looking at the grammar generated by interaction terms in the Standard Model. The nonterminal symbols will be the letters S (the head symbol), A and B. The terminal symbols will be the letters x and y. The production rules will be

$$S \rightarrow AB \qquad \text{Rule I}$$
$$A \rightarrow y \qquad \text{Rule II}$$
$$A \rightarrow Ay \qquad \text{Rule III}$$
$$B \rightarrow x \qquad \text{Rule IV}$$
$$B \rightarrow Bx \qquad \text{Rule V}$$

The Chomsky computer language that this grammar generates consists of all strings containing any number of y's followed by any number of x's since any of these strings can be generated from the head symbol S using the production rules. Because of rule I the y's are placed to the left and x's are placed to the right. The order of the symbols matters just as it does in human language – consider the words ides and dies which differ only in the order of i and d! An example of generating a string yyxxx from the head symbol is:

$$S \qquad \rightarrow AB \qquad \text{Rule I}$$

$$
\begin{aligned}
AB &\rightarrow AyB & \text{Rule III}\\
AyB &\rightarrow yyB & \text{Rule II}\\
yyB &\rightarrow yyBx & \text{Rule V}\\
yyBx &\rightarrow yyBxx & \text{Rule V}\\
yyBxx &\rightarrow yyxxx & \text{Rule IV}
\end{aligned}
$$

The production rule used to make each transition is listed above on the right.

Our generalization of the Chomsky definition of language would allow any string to be the starting point – not just the head symbol S. Using the sample grammar described on the previous page the generalized language becomes any string of x's and y's.

A more interesting language can be created by adding two new rules to the rules on the preceding page:

$$
\begin{aligned}
y &\rightarrow A & \text{Rule VI}\\
x &\rightarrow B & \text{Rule VII}
\end{aligned}
$$

The resulting language – the set of strings of terminal symbols – remains the same despite the addition of these new grammar rules. However the number and variety of transitions becomes much larger. For example the following chain of transitions is allowed,

$$ yx \rightarrow AB \rightarrow AyBx \rightarrow AyyBx \rightarrow AyyyBxx \rightarrow yyyyxx $$

In the next section we will extend the concept of computer grammars by allowing probabilistic grammar rules – production rules which have an associated probability of executing.

9-A.4 Probabilistic Computer Grammars

> *Grammar, which knows how to control even kings.*
> *Molière - Les Femmes Savantes (1672) Act II, Scene 6*

9-A.4.1 Probabilistic Computer Grammars™

The preceding section described the production rules for a *deterministic grammar*. The left side of each production rule has one, and only one, possible transition.

Non-deterministic grammars allow two or more grammar rules to have the same left side, and different right sides. For example,

$$
\begin{aligned}
A &\rightarrow y\\
A &\rightarrow x
\end{aligned}
$$

could both appear in a non-deterministic grammar.

Non-deterministic grammars can be easily (almost "naturally") associated with probabilities. The probabilities can be classical probabilities or quantum probabilities. An example of a simple non-deterministic grammar is specified by the production rules:

$$S \rightarrow xy \quad \textbf{Rule I}$$
$$x \rightarrow xx \quad \textbf{Rule II} \quad \textbf{Relative Probability} = .75$$
$$x \rightarrow xy \quad \textbf{Rule III} \quad \textbf{Relative Probability} = .25$$
$$y \rightarrow yy \quad \textbf{Rule IV}$$

where the head symbol is the letter S, and the terminal symbols are the letters x and y. The relative probability of generating the string xxy vs. the relative probability of generating the string xyy from the string xy is

$$xy \rightarrow xxy \qquad \text{relative probability} = .75$$
$$xy \rightarrow xyy \qquad \text{relative probability} = .25$$

The string xxy is three times more likely to be produced than the string xyy.

For each starting string one can obtain the relative probabilities that various possible output strings will be produced.

A more practical example of a Probabilistic Grammar™ can be abstracted from flipping coins – heads or tails occur with equal probability – 50-50. From this observation we can create a little Probabilistic Grammar™ for the case of flipping two coins. Let h represent heads and t represent tails. Then consider the grammar:

$$S \rightarrow hh$$
$$S \rightarrow tt$$
$$S \rightarrow ht$$
$$S \rightarrow th$$
$$h \rightarrow t \qquad \text{probability} = .5 \ (50\%)$$
$$h \rightarrow h \qquad \text{probability} = .5 \ (50\%)$$
$$t \rightarrow h \qquad \text{probability} = .5 \ (50\%)$$
$$t \rightarrow t \qquad \text{probability} = .5 \ (50\%)$$

The last four rules above embody the statement that flipping a coin yields heads or tails with equal probability (50% or .5).

Now let us consider starting with two heads hh. The possible outcomes and their probabilities are:

$$hh \rightarrow hh \qquad \text{probability} = .5 * .5 = .25$$
$$hh \rightarrow th \qquad \text{probability} = .5 * .5 = .25$$
$$hh \rightarrow ht \qquad \text{probability} = .5 * .5 = .25$$

$$hh \rightarrow tt \quad \text{probability} = .5 * .5 = .25$$

If we don't care about the order of the output heads and tails, then the probability of flipping two heads and getting a head and tail (hh → ht or hh → th) is .25 + .25 = .5.

This simple example shows the basic thought process of a non-deterministic grammar with associated probabilities.

The combination of a non-deterministic grammar and an associated set of probabilities for transitions can be called a *Probabilistic Grammar*. We will see that the grammar production rules for the Standard Model must be viewed as constituting a Probabilistic Grammar with one difference. The "square roots" of probabilities – probability amplitudes – are specified for the transitions in the grammar. The Standard Model requires probability amplitudes since it is a quantum theory. Therefore we will describe probabilistic grammars with associated probability amplitudes (such as that of the Standard Model) as *Quantum Probabilistic Grammars*.

9-A.4.2 Quantum Probabilistic Grammar

An example of a Quantum Probabilistic Grammar can be constructed based on an analogy with a two slit photon experiment. Imagine a wall with two slits. A source shoots photons at the wall. A photon can go through either slit with equal quantum probability. An illustration of this experimental arrangement is:

Figure 9-A.3. Two slit photon experimental setup.

A simple Quantum Probabilistic Grammar can be constructed corresponding to this experimental setup:

$$S \rightarrow 1 \quad \text{probability amplitude} = 1/\sqrt{2}$$
$$S \rightarrow 2 \quad \text{probability amplitude} = 1/\sqrt{2}$$

The head symbol S represents the source. The digit 1 represents a photon going through slit 1. The digit 2 represents a photon going through slit 2.

The values of the probability amplitudes $1/\sqrt{2}$ can be calculated using Quantum Mechanics. The probability for a photon to go through slit 1 is the absolute value squared of the probability amplitude:

$$\text{Probability to go through slit } 1 = (1/\sqrt{2})^2 = .5$$

and the probability for a photon to go through slit 2 is

$$\text{Probability to go through slit } 2 = (1/\sqrt{2})^2 = .5$$

This simple example illustrates the basics of a Quantum Probabilistic Grammar.

Before applying these concepts to the Standard Model we will look at a simpler Quantum Field Theory called a ϕ^3 ("phi cubed") theory (ϕ is the Greek letter phi). This theory describes a self-interacting spin 0 particle with no internal symmetries. This theory is a stepping stone to far more complex Standard Model Quantum Field Theories. We are only interested in it as a simple example of quantum probabilistic grammar rules.

The ϕ^3 theory is so named because it has a cubic Lagrangian interaction term. (Note the exponent 3.) The grammar rules for the ϕ^3 theory are:

$$\phi \rightarrow \phi\phi \qquad\qquad \text{Rule I}$$
$$\phi\phi \rightarrow \phi \qquad\qquad \text{Rule II}$$

Rule I corresponds to the emission of a ϕ particle and rule II corresponds to the absorption of a ϕ particle. We will not introduce a start symbol. Instead we will consider the transitions from an input state of a number of ϕ particles to an output state of (possibly) a different number of ϕ particles. *We will ignore the momenta of the particles.* (This assumption is equivalent to assuming the ϕ particles have infinite mass.)

We will assume either transition above takes place with a "relative probability amplitude" g. We will call this simplified theory the *modified ϕ^3 theory*. We will view g as a measure of the probability amplitude for an absorption or emission of a ϕ particle. (g is similar to a coupling constant in Quantum Field Theory.) The probabilities have to be normalized or rescaled so that the sum of all probabilities equals one.

To get a feel for the Quantum Probabilistic Grammar approach we will look at the case of an input state consisting of two ϕ particles. The output states can have one ϕ particle, two ϕ's, three ϕ's, and so on. Each possible output state has a certain probability of occurring. The sum of the probabilities for producing all possible output states must equal one. (Remember that the sum of all possible outcomes of flipping a coin is one. Having it come up heads has probability ½ and having it come up as a tails has probability ½ also.)

The simplest string transition from a two ϕ "input" state to a one ϕ "output" state is:

$$\phi\phi \to \phi$$

using Rule II. The probability amplitude of this transition is g by assumption.

The transitions between strings can be visualized with diagrams that are like the Feynman diagrams that used in Quantum Field Theory perturbation theory calculations. These diagrams are not the same as Feynman diagrams because they embody time orderings of emissions and absorptions of ϕ particles. (They actually harken back to the time-ordered diagrams that were used by physicists prior to 1950.) **This feature supports the need for the Principle of Asynchronicity described earlier.**

In some simple cases the time ordering is irrelevant. For example, the Feynman-like diagram for the simplest case of a two ϕ input state transitioning *directly* to a one ϕ output state is the same as the Feynman diagram:

Figure 9-A.4. Diagram for $\phi\phi \to \phi$. The input states are always on the left and the output states are always on the right in Feynman and Feynman-like diagrams.

The time order of emission and absorption of ϕ particles can be symbolized using parentheses. For example,

$$(\phi)\phi \to (\phi\phi)\phi = \phi(\phi\phi) \to \phi(\phi) = \phi\phi \qquad \text{Diagram A (Fig. 9-A.5)}$$

and

$$\phi(\phi) \to \phi(\phi\phi) = (\phi\phi)\phi \to (\phi)\phi = \phi\phi \qquad \text{Diagram B (Fig. 9-A.5)}$$

These string transitions correspond to different time-ordered Feynman-like diagrams:

A　　　　　　B

Feynman diagram Feynman-like diagrams with different time orderings

Figure 9-A.5 The true Feynman diagram is the "sum" of the two time-ordered Feynman-like diagrams.

The correspondence between Feynman-like diagrams and the transitions between strings based on the Quantum Probabilistic Grammar can be seen by taking vertical slices on diagrams A or B above after each emission or absorption. For example,

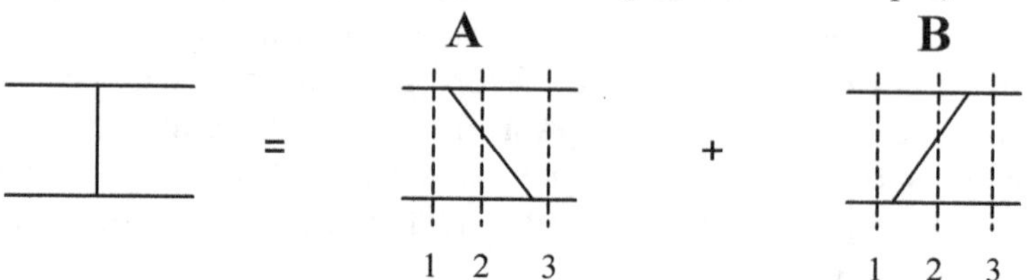

Figure 9-A.6 Slicing Feynman-like diagrams. A slice is made after each emission or absorption. As you read down a slice the particles are listed in the same order as the corresponding string. Each slice is numbered starting from the left.

The string corresponding to each numbered slice in the above figures is similarly numbered in the following transitions:

Slice:

$$1 \qquad\qquad 2 \qquad\qquad 3$$
$$(\phi)\phi \quad \rightarrow \quad (\phi\phi)\phi = \phi(\phi\phi) \quad \rightarrow \quad \phi(\phi) = \phi\phi \; \mathbf{A}$$

Slice:

$$1 \qquad\qquad 2 \qquad\qquad 3$$
$$\phi(\phi) \quad \rightarrow \quad \phi(\phi\phi) = (\phi\phi)\phi \quad \rightarrow \quad (\phi)\phi = \phi\phi \; \mathbf{B}$$

Parentheses on the left side of an arrow indicate the particle(s) that emits a new particle(s) appearing within the corresponding parentheses on the right side of the arrow.

A transition from an input state containing ϕ particles to an output state containing ϕ particles always has an infinite number of ways of taking place and thus an infinite number of Feynman-like diagrams. Readers familiar with the perturbation theory of Quantum Field Theory will remember that these diagrams are the same as the Feynman diagrams generated by perturbation theory with the additional feature of having time orderings.

We will now look at the transition of two ϕ particles to two ϕ particles: $\phi\phi \rightarrow \phi\phi$. There are an infinite number of Feynman-like diagrams for this transition. Some of the simpler Feynman diagrams are:

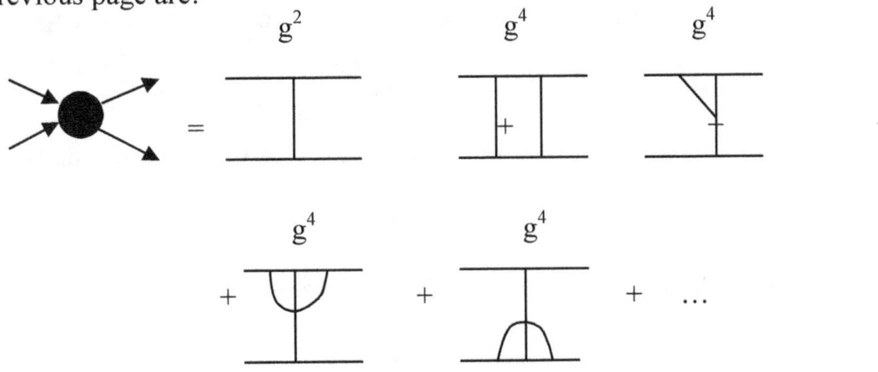

Figure 9-A.7. Diagrams for an elastic collision with two incoming particles and two outgoing particles.

Each Feynman diagrams corresponds to several time-ordered Feynman-like diagrams.

In evaluating these diagrams to calculate probabilities we must remember that we are ignoring space-time aspects such as particle propagators and momenta. So the calculation of the probability amplitude for this process becomes a counting problem of the number of diagrams that exist for each power of g^2. The probability amplitude for each diagram is a power of g^2.

Counting diagrams is a combinatorial mathematics problem that we will not explore in detail because it is peripheral to our interests. Consequently we will simply express the probability amplitude as:

$$A_2(g) = \sum_{n=1}^{\infty} a_n\, g^{2n}$$

where the mathematical expression on the right represents a sum from $n = 1$ to infinity and where the numbers a_n are integer numbers equal to the number of different diagrams having a power of g^{2n} as its probability amplitude. Each intersection of lines (called a vertex) contributes a factor of g to the amplitude for that diagram. The powers of g for the simpler diagrams that appear on the previous page are:

Figure 9-A.8. The power of g for some simple diagrams.

The value of the first constant a_1 is 1 since there is only one Feynman diagram with amplitude value g^2 for this process – the first diagram to the right of the = sign in Fig. 9-A.8. We treat all time-ordered variations of a Feynman diagram as contributing one to the value of a_n. The value of a_n grows rapidly as n increases. For large n the value of a_n is of the order of $(n!)^2$. Consequently the sum is an asymptotic power series. Here again the details are not important for us. We are not looking for a numerical result.

The (unnormalized) relative probability for the transition $\phi\phi \rightarrow \phi\phi$ is:

$$P_2 = |A_2(g)|^2$$

where $|...|$ represents the absolute value of the complex probability amplitude. In quantum theories a probability is always the square of the absolute value of its probability amplitude. The probability P_2 is a relative probability that must be normalized – multiplied by a factor that makes the sum of the probabilities of all possible outcome states equal to one. To calculate this probability we must calculate the sum of all the relative probabilities P_n to produce any number of ϕ particles from a two ϕ input state.

$$\phi\phi \rightarrow \phi \,...$$

The total of the relative probabilities is:

$$P = \sum_{n=1}^{\infty} P_n$$

where P_n is the relative probability to produce an output state with n ϕ particles.

The calculation of the relative probabilities P_n for n ϕ particles output states is similar to the calculation P_2. For example, for three particles

$$A_3(g) = \sum_{n=1}^{\infty} b_n\, g^{2n+1}$$

where the numbers b_n count the number of distinct diagrams with the power g^{2n+1} and

$$P_3 = |A_3(g)|^2$$

The absolute (normalized) probability to produce an n ϕ particle output state is

$$Q_n = P_n/P$$

The sum of all possible output state probabilities equals one:

$$1 = \sum_{n=1}^{\infty} Q_n$$

The modified ϕ^3 Quantum Field Theory provides a simple example of a Quantum Probabilistic Grammar. We will now turn to the Standard Model and examine its Quantum Probabilistic Grammar. Because it encompasses a much larger number of different particles (letters) and interactions (grammar rules) it will be significantly more complex.

9-A.4.3 Probability Amplitudes of Quantum Grammars

Quantum Grammar rules associate a probability amplitude with each production rule. To find the probability for a transition from an initial state to a final state we must calculate the probability amplitude for the transition through the repeated application of the production rules for each possible path from the initial state to the final state. We assume each initial state of the Quantum Turing machine begins with probability amplitude one. (This is a normalization condition for the initial state in reality.)

When a production rule is applied to a state to produce a transition to a new state the current probability amplitude is multiplied by the probability amplitude of the production rule. Thus the total relative probability for the passage from an initial state i to a specific final state f is

$$P_{fi} = \left| \sum_{\text{paths}} a_1 \, a_2 \, a_3 \, \ldots \, a_{n(\text{path})} \right|^2$$

where the sum is over all finite paths that lead from the initial state to the final state through the application of all relevant production rules, and where $a_1 a_2 \, a_3 \, \ldots \, a_{n(\text{path})}$ is the product of the probability amplitudes of the production rules for each individual path. P_{fi} is similar in form to a Feynman path integral expression (See Feynman (1965)).

The value of n is path dependent and thus denoted n(path). The relative probability is the absolute value squared of the sum of products of the amplitudes. The relative probability must be normalized to produce an absolute probability such that the sum of the absolute probabilities of all possible final states is one:

$$P_{\text{absolute-}fi} = NP_{fi}$$

$$N = \sum_{f} P_{fi}$$

$$1 = \sum_{f} P_{\text{absolute-}fi}$$

where the sums are over all possible final states f.

Thus we have a well-defined method for calculating the probability of a transition from an initial state to a final state that is illustrated by the preceding examples. The fact that it bears some resemblance to the path integral methods for quantum mechanics pioneered by Feynman suggests that Quantum Grammars are of interest to physics. This author was first struck by the similarity of string transitions via production rules to path integrals in 1981. After all, a jagged

(discrete) Feynman path is really a string of coordinates marking the end points of each line segment of which the path is composed. Thus each path in a Feynman sum over paths can be represented by a string. The evolution of a path from line segment to line segment can be viewed as the repeated application of a probabilistic production rule. The path sum equivalent of the probability amplitude of a production rule is an exponential Hamiltonian factor that is a function of the change in string coordinates "due to the production rule."

9-A.5 Standard Model Quantum Grammar

9-A.5.1 Grammar Production Rules of Quantum Electrodynamics

We now consider the grammar production rules of the Quantum Electrodynamics (electromagnetism) sector of the Standard Model. The production rules corresponding to the electromagnetic interaction term for electrons and positrons in the Standard Model

$$\bar{e}Ae$$

are:

ELECTRON-POSITRON QED PRODUCTION RULES

$$e \rightarrow eA$$
$$e \rightarrow Ae$$
$$eA \rightarrow e$$
$$Ae \rightarrow e$$
$$p \rightarrow pA$$
$$p \rightarrow Ap$$
$$Ap \rightarrow p$$
$$pA \rightarrow p$$
$$ep \rightarrow A$$
$$pe \rightarrow A$$
$$A \rightarrow ep$$
$$A \rightarrow pe$$

where e represents an electron, p represents a positron, and A represents a photon. The production rules describe the emission and absorption of photons by electrons and positrons as well as the annihilation of an electron and positron to produce a photon, and the decay of a photon into an electron-positron pair.

An example of an interaction between two electrons in the linguistic approach is:

$$\begin{array}{ccccc} 1 & & 2 & & 3 \\ ee & \rightarrow & eAe & \rightarrow & ee \end{array}$$

where the electrons interact by exchanging one photon. One Feynman-like diagram for these transitions is:

Figure 9-A.9. A diagram showing how two electrons interact by exchanging a photon. As time increases the electrons move from left to right. The upper electron emits the photon. This corresponds to the left e transitioning to eA using the grammar rule e → eA. (There is a similar diagram Fig. 9-A.10 in which the lower electron emits the photon.)

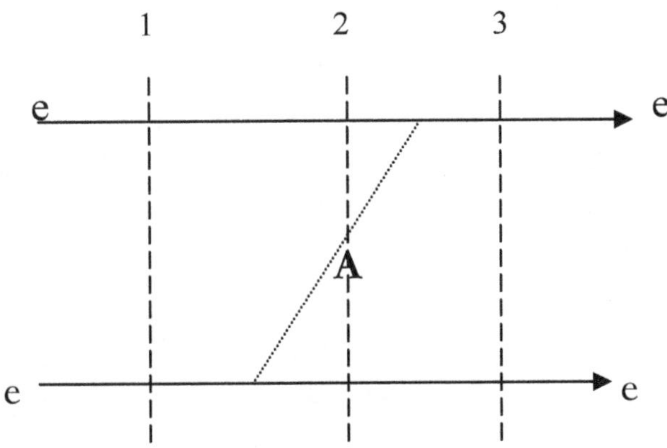

Figure 9-A.10. Another diagram with two electrons interacting by exchanging a photon. In this case the lower electron emits the photon. This corresponds to the right e in the initial ee string transitioning to Ae using the grammar rule e → Ae.

The vertical slices in the Feynman diagram which are numbered 1, 2 and 3 correspond to the three numbered strings in the transitions generated from the production rules. Each string has an ordering that corresponds to the order of particles as you descend a slice. For example slice 2 in Fig. 9-A.10 has an electron, photon, and another electron in that order as you descend corresponding to string 2 above.

 Another Feynman-like diagram that contributes to this process has the lower electron emitting a photon that is then absorbed by the upper electron. The Feynman diagram for this process represents the sum of both of the previous diagrams:

Feynman diagram Feynman-like diagrams with different time orderings

Figure 9-A.11. A Feynman diagram represents several of our Feynman-like diagrams with different time orderings of particle emission and absorption.

A more complex example of a Feynman-like diagram appears in Fig. 9-A.12. Six slices appear corresponding to the various intermediate states in this complex electron-electron interaction. The production rules can be used to generate a sequence of strings that correspond to the slices. As you descend each slice the particles are ordered in the same way as the corresponding string. For example, as you descend slice 5 the order of the particles is electron, electron, positron, photon and electron, and the string is eepAe.

$$\begin{array}{cccccc} 1 & 2 & 3 & 4 & 5 & 6 \end{array}$$

$$ee \;\rightarrow\; eAAe \rightarrow eAe \rightarrow\; eAAe \;\rightarrow\; eepAe \;\rightarrow\; eepe$$

Figure 9-A.12. A diagram for the collision of two electrons that produce a new electron-positron pair: ee → eepe.

The transitions between character strings have an ambiguity. For example in the above transition

$$\begin{array}{ccc} eAAe & \rightarrow & eAe \\ 2 & & 3 \end{array}$$

could have taken place through (eA)Ae → (e)Ae with the left (upper) e absorbing an A or through eA(Ae) → eA(e) with the right (lower) e absorbing an A. (Parentheses are used for grouping to show which electron absorbed the photon.) This ambiguity reflects the fact that there are several possible time orderings. The preceding diagram actually corresponds to eA(Ae) → eA(e). The right (lower) electron absorbs the photon.

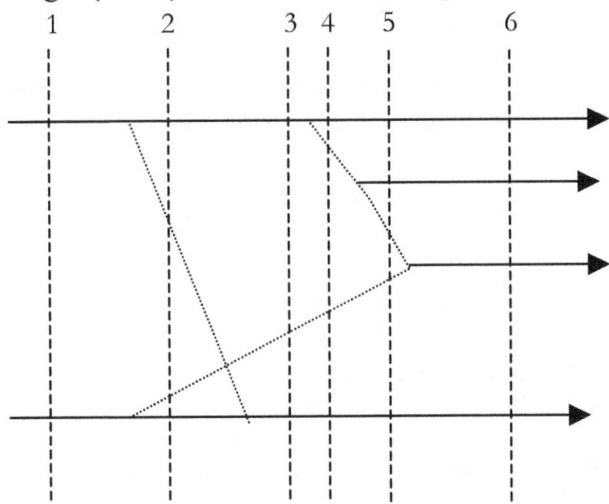

To calculate the probability of an actual physical transition occurring such as ee → eepe we must take account of all diagrams with all possible time orderings for the specified input and output states. This is an impossiblel chore since an infinite number of diagrams are involved. Normally only the simplest diagrams are evaluated since they dominate the electromagnetic interactions of electrons and positrons.

The above examples show how the electromagnetic interaction part of the Standard Model lagrangian can be viewed as defining a grammar. The grammar has corresponding

Feynman-like diagrams. As we pointed out earlier, these types of diagrams have time orderings that are similar to the time orderings that appeared in perturbation theory calculations before 1950.

The Weak Interaction and Strong Interaction parts of the Standard Models also define grammars. Consequently we can view a complete Standard Model Lagrangian as defining a grammar where the "letters" (alphabet or vocabulary) are the elementary particles of the model and the Feynman-like diagrams corresponding to the Standard Model are a sequence of strings generated by applying the production rules specified by the Standard Model Lagrangian.

9-A.5.2 Production Rules for the Weak and Strong Interactions

The Weak and Strong interaction terms in the Standard Model Lagrangian are also easily translated into grammar production rules (although the process is laborious since there are so many of them). We will illustrate these cases using the Weak interaction terms:

$$\bar{\nu}_e W^- e$$

and

$$\bar{\nu}_\mu W^- \mu$$

where ν_e represents an electron neutrino, ν_μ represents a muon neutrino, μ represents a muon, W^- is a gauge field of the Weak interaction and e is an electron; and the Strong interaction term

$$\bar{u} G u$$

where u is a u quark and G represents gauge fields of the Strong interaction.

Notice that there are several types of neutrinos: electron neutrinos, muon neutrinos and tau neutrinos. The three kinds of neutrinos have different internal quantum numbers that distinguish them. Neutrinos do not have electromagnetic charge. They are neutral as their name suggests. Each kind of neutrino has a corresponding charged partner. We are familiar with the electron. The other charged partners are the muon and tau particle. These charged particles are like heavy electrons for the most part. The three charged leptons also have distinguishing internal quantum numbers.

The preceding interaction terms imply production rules such as:

$$e \rightarrow W^- \nu_e$$

$$e \rightarrow \nu_e W^-$$

$$W^- \rightarrow e \nu_e$$

$$W^- \rightarrow \nu_e e$$

$$\mu \rightarrow \nu_\mu\, W^-$$

$$\mu \rightarrow W^-\nu_\mu$$

$$u \rightarrow Gu'$$

$$u \rightarrow u'G$$

and so on where e is an electron, p is a positron, W^- is a negative W gauge boson, ν is a neutrino, G is a Strong interaction gauge boson and u and u' are u quarks which may have different color quantum numbers.

These production rules generate string equivalents of Weak interaction transitions such as muon decay:

$$\begin{array}{cccc} 1 & 2 & 3 \\ \mu & \rightarrow\ \ W^-\nu_\mu & \rightarrow\ \ e\nu_e\nu_\mu \end{array}$$

which has the corresponding Feynman-like diagram:

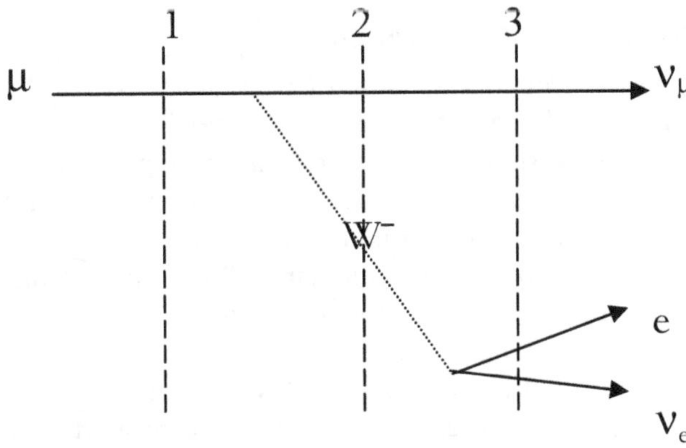

They also generate string equivalents of Strong interaction transitions between quarks and color gauge fields such as:

$$\begin{array}{ccc} 1 & 2 & 3 \\ uu & \rightarrow uGu & \rightarrow u'u' \end{array}$$

with the corresponding Feynman-like diagram:

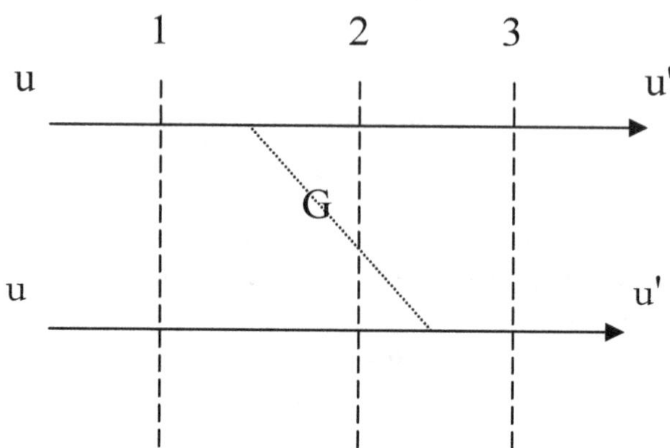

The preceding examples are among the simplest cases of the infinite variety of Feynman-like diagrams that can be generated from the Standard Model production rules.

9-A.5.3 A Standard Model Quantum Grammar

At this point we have created a view of a Standard Model of elementary particles in which the particles are an alphabet of 'letters.' The particle alphabet can be combined into strings that represent input or output states of scattering particles as well as bound states. Transitions between strings take place through quantum grammar production rules and correspond to time-ordered Feynman-like diagrams.

So we now have a map between a Standard Model and a language, with letters, words and a grammar; and an interpretation of the language in terms of rules of calculation and experimental setups.

The linguistic representation of a Standard Model that we have developed omits many important calculation details as well as important particle properties such as spin and particle momenta. These features could be added in a direct way. The focus of our investigation is on the essentials of the acts of creation and annihilation of particles in particle interactions.

The character string transition approach based on production rules is equivalent to the Feynman diagram approach. It does however provide a different, and simpler, view. Interestingly the Feynman diagram approach to calculations is very difficult and tedious – but it often leads to a simple result due to the massive cancellation of many complex terms with each other. (An attempt to simplify perturbation theory diagram calculations was made by Cheng and Wu in the early 1970's and by this author privately.) The simple linguistic approach might be a hint of a more efficient way of calculating in quantum field theories like the Standard Model where complications are absent from the very beginning.

Whether or not the linguistic approach leads to a less complicated theory or method of calculation remains to be seen. However we have now obtained a rather amazing result. After

2500 years of speculation on the nature of matter we have developed a surprisingly simple theory (for everything except gravitation) called the Standard Model that can be viewed, in part, as a quantum type 0 computer language. It has an alphabet (vocabulary) of particles, and a set of production rules specified by the interaction part of the Standard Model Lagrangian.

This situation represents something of a miracle. There is no reason that Nature should have so few particles that interact with each other through a simple set of rules. A computer language theorist would call the language of the Standard Model a language with a finite representation. Simply put, this means the words of the language can be generated from a finite vocabulary (alphabet or set of particles) and a finite set of production rules.

9-A.5.4 The Standard Model Language is Surprisingly Simple

Many physicists have felt that there are too many elementary particles in the Standard Model. From the point of view of a language theorist *the finite language representation of the Standard Model is a very special situation.* As Hopcroft and Ullman[93] point out, "there are many more languages than finite representations." Languages can have infinite alphabets or infinite sets of production rules or other complications.

The physical equivalent of an infinite alphabet would be a universe with an infinite number of different types of matter. Every particle of matter could have a different mass and differ in other properties. From this point of view the Standard Model is truly a marvel of simplicity.

The Standard Model, in fact, is a very compact, finite description of most of the known features of our universe. The linguistic view of the Standard Model suggests we should view elementary particles as symbols or clumps of data – a vocabulary. The interactions of the elementary particles involve the creation or annihilation of particles – creation in the deepest form seen by man.

Remarkably, the data flowing through a computer can be viewed as being transformed from one form to another and being output to different destinations. Data flowing through a computer can be divided into streams that can be sent to different output channels such as a printer or the computer screen. For example, the character, 'a', can be a data item in a stream of data that is sent to both a printer and the screen:

The streaming of data characters to different output channels is quite analogous to the output of particles from particle scattering as we have seen.

[93] J. E. Hopcroft and J. D. Ullman, *Formal Languages and Their Relation to Automata*, (Addison-Wesley, Reading, MA, 1975) page 2.

The simple production rules that describe particle scattering in the Standard Model suggest a fundamental simplicity at the core of Reality that goes far beyond the speculations of philosophers and scientists of earlier ages.

10. The Unmoved Mover of Physics

My cathedral is the universe;
My candles the stars.

We have developed a unified SuperStandard Model that includes Gravitation and the known particles and interactions of Nature. Implcit in this theory is the assumption that the 'real world' corresponds to our model (or perhaps a better model if developed in future) in every detail for all time.

Yet philosophers and scientists have wondered, since at least the time of the Pre-Socratic philosophers,[94] why our thoughts (theories) conform to reality. For it is one thing to theorize about reality, and yet another thing to confront reality. *The link between the mathematics of physical theory and material reality has not been found.*

Why the harmony of thought and reality? What coordinates the two? What originated reality in the manner of our theories? What moves reality to evolve in accord with our theories down to the deepest possible level?

There are two major issues related to these questions:

1. What is the source of the equations and constructs that define the physical nature of reality?

2. What causes reality to conform to mathematical law? Or, put differently, why does reality specify its laws of behavior mathematically as we have discovered over the millennia?

Ancient philosophers such as Aristotle developed the concept of the 'Unmoved Mover' to answer these questions. This entity originated and caused the movement of all in the universe. It was the ultimate motivator. Yet it itself did not change. It was 'unmoved.'

There seems to be an entity like the 'Unmoved Mover.' Otherwise one only has the possibilities of a chaotic universe or of a 'self-organizing' universe. But we do not see chaos. And a self-organizing universe can only be so if local deviations from physical law exist, here and there, in the universe. These possible deviations have not been observed.

Thus we conclude these two alternatives do not stand up to scrutiny. Nor do suggestions of a beginningless-endless universe, perhaps oscillating, that merely put off the question.

The only sensible possibility is the existence of a Supreme Being – God – the Unmoved Mover. A God –infinite – existing everywhere in the universe – causing all to move in accord

[94] See Kirk (1962).

with His plan which happens to be mathematical in nature – and which we call Physics. No other alternative seems possible as an explanation of the success of Science.

This concept has been called the ultimate proof of the existence of God for over twenty five hundred years. Our work in volume I, Blaha (2017c), and this book stands as evidence of this.

Science is one of Mankind's ways of revealing God's Plan.

REFERENCES

Akhiezer, N. I., Frink, A. H. (tr), 1962, *The Calculus of Variations* (Blaisdell Publishing, New York, 1962).

Bjorken, J. D., Drell, S. D., 1964, *Relativistic Quantum Mechanics* (McGraw-Hill, New York, 1965).

Bjorken, J. D., Drell, S. D., 1965, *Relativistic Quantum Fields* (McGraw-Hill, New York, 1965).

Blaha, S., 1998, *Cosmos and Consciousness* (Pingree-Hill Publishing, Auburn, NH, 1998).

______, 2002, *A Finite Unified Quantum Field Theory of the Elementary Particle Standard Model and Quantum Gravity Based on New Quantum Dimensions™ & a New Paradigm in the Calculus of Variations* (Pingree-Hill Publishing, Auburn, NH, 2002).

______, 2003, *A Finite Unified Quantum Field Theory of the Elementary Particle Standard Model and Quantum Gravity Based on New Quantum Dimensions™ and a New Paradigm in the Calculus of Variations* (Pingree-Hill Publishing, Auburn, NH, 2003).

______, 2004, *Quantum Big Bang Cosmology: Complex Space-time General Relativity, Quantum Coordinates™Dodecahedral Universe, Inflation, and New Spin 0, ½, 1 & 2 Tachyons & Imagyons* (Pingree-Hill Publishing, Auburn, NH, 2004).

______, 2005a, *Quantum Theory of the Third Kind: A New Type of Divergence-free Quantum Field Theory Supporting a Unified Standard Model of Elementary Particles and Quantum Gravity based on a New Method in the Calculus of Variations* (Pingree-Hill Publishing, Auburn, NH, 2005).

______, 2005b, *The Metatheory of Physics Theories, and the Theory of Everything as a Quantum Computer Language* (Pingree-Hill Publishing, Auburn, NH, 2005).

______, 2005c, *The Equivalence of Elementary Particle Theories and Computer Languages: Quantum Computers, Turing Machines, Standard Model, Superstring Theory, and a Proof that Gödel's Theorem Implies Nature Must Be Quantum* (Pingree-Hill Publishing, Auburn, NH, 2005).

______, 2006a, *The Foundation of the Forces of Nature* (Pingree-Hill Publishing, Auburn, NH, 2006).

______, 2006b, *A Derivation of ElectroWeak Theory based on an Extension of Special Relativity; Black Hole Tachyons; & Tachyons of Any Spin.* (Pingree-Hill Publishing, Auburn, NH, 2006).

______, 2007a, *Physics Beyond the Light Barrier: The Source of Parity Violation, Tachyons, and A Derivation of Standard Model Features* (Pingree-Hill Publishing, Auburn, NH, 2007).

______, 2007b, *The Origin of the Standard Model: The Genesis of Four Quark and Lepton Species, Parity Violation, the ElectroWeak Sector, Color SU(3), Three Visible Generations of Fermions, and One Generation of Dark Matter with Dark Energy* (Pingree-Hill Publishing, Auburn, NH, 2007).

______, 2008a, *A Direct Derivation of the Form of the Standard Model From GL(16) (Pingree-Hill Publishing, Auburn, NH, 2008).*

______, 2008b, *A Complete Derivation of the Form of the Standard Model With a New Method to Generate Particle Masses Second Edition* (Pingree-Hill Publishing, Auburn, NH, 2008)

______, 2009, *The Algebra of Thought & Reality: The Mathematical Basis for Plato's Theory of Ideas, and Reality Extended to Include A Priori Observers and Space-Time Second Edition* (Pingree-Hill Publishing, Auburn, NH, 2009).

______, 2010a, *Operator Metaphysics: A New Metaphysics Based on a New Operator Logic and a New Quantum Operator Logic that Lead to a Mathematical Basis for Plato's Theory of Ideas and Reality* (Pingree-Hill Publishing, Auburn, NH, 2010).

______, 2010b, *The Standard Model's Form Derived from Operator Logic, Superluminal Transformations and GL(16)* (Pingree-Hill Publishing, Auburn, NH, 2010).

______, 2010c, *SuperCivilizations: Civilizations as Superorganisms* (McMann-Fisher Publishing, Auburn, NH, 2010).

______, 2011a, *21st Century Natural Philosophy Of Ultimate Physical Reality* (McMann-Fisher Publishing, Auburn, NH, 2011).

______, 2011b, *All the Universe! Faster Than Light Tachyon Quark Starships & Particle Accelerators with the LHC as a Prototype Starship Drive Scientific Edition* (Pingree-Hill Publishing, Auburn, NH, 2011).

______, 2011c, *From Asynchronous Logic to The Standard Model to Superflight to the Stars* (Blaha Research, Auburn, NH, 2011).

______, 2012a, *From Asynchronous Logic to The Standard Model to Superflight to the Stars volume 2: Superluminal CP and CPT, U(4) Complex General Relativity and The Standard Model, Complex Vierbein General Relativity, Kinetic Theory, Thermodynamics* (Blaha Research, Auburn, NH, 2012).

______, 2012b, *Standard Model Symmetries, And Four And Sixteen Dimension Complex Relativity; The Origin Of Higgs Mass Terms* (Blaha Reasearch, Auburn, NH, 2012).

______, 2013a, *Multi-Stage Space Guns, Micro-Pulse Nuclear Rockets, and Faster-Than-Light Quark-Gluon Ion Drive Starships* (Blaha Research, Auburn, NH, 2013).

______, 2013b, *The Bridge to Dark Matter; A New Sister Universe; Dark Energy; Inflatons; Quantum Big Bang; Superluminal Physics; An Extended Standard Model Based on Geometry* (Blaha Reasearch, Auburn, NH, 2013).

______, 2014a, *Universes and Megaverses: From a New Standard Model to a Physical Megaverse; The Big Bang; Our Sister Universe's Wormhole; Origin of the Cosmological Constant, Spatial Asymmetry of the Universe, and its Web of Galaxies; A Baryonic Field between Universes and Particles; Megaverse Extended Wheeler-DeWitt Equation* (Blaha Reasearch, Auburn, NH, 2014).

______, 2014b, *All the Megaverse! Starships Exploring the Endless Universes of the Cosmos Using the Baryonic Force* (Blaha Research, Auburn, NH, 2014).

______, 2014c, *All the Megaverse! II Between Megaverse Universes: Quantum Entanglement Explained by the Megaverse Coherent Baryonic Radiation Devices – PHASERs Neutron Star Megaverse Slingshot Dynamics Spiritual and UFO Events, and the Megaverse Microscopic Entry into the Megaverse* (Blaha Research, Auburn, NH, 2014).

______, 2015a, *PHYSICS IS LOGIC PAINTED ON THE VOID: Origin of Bare Masses and The Standard Model in Logic, U(4) Origin of the Generations, Normal and Dark Baryonic Forces, Dark Matter, Dark Energy, The Big Bang, Complex General Relativity, A Megaverse of Universe Particles* (Blaha Research, Auburn, NH, 2015).

______, 2015b, *PHYSICS IS LOGIC Part II: The Theory of Everything, The Megaverse Theory of Everything, U(4)⊗U(4) Grand Unified Theory (GUT), Inertial Mass = Gravitational Mass, Unified Extended Standard Model and a New Complex General Relativity with Higgs Particles, Generation Group Higgs Particles* (Blaha Research, Auburn, NH, 2015).

______, 2015c, *The Origin of Higgs ("God") Particles and the Higgs Mechanism: Physics is Logic III, Beyond Higgs – A Revamped Theory With a Local Arrow of Time, The Theory of Everything Enhanced, Why Inertial Frames are Special, Universes of the Mind* (Blaha Research, Auburn, NH, 2015).

______, 2015d, *The Origin of the Eight Coupling Constants of The Theory of Everything: U(8) Grand Unified Theory of Everything (GUTE), S^8 Coupling Constant Symmetry, Space-Time Dependent Coupling Constants, Big Bang Vacuum Coupling Constants, Physics is Logic IV* (Blaha Research, Auburn, NH, 2015).

______, 2016a, *New Types of Dark Matter, Big Bang Equipartition, and A New U(4) Symmetry in the Theory of Everything: Equipartition Principle for Fermions, Matter is 83.33% Dark, Penetrating the Veil of the Big Bang, Explicit QFT Quark Confinement and Charmonium, Physics is Logic V* (Blaha Research, Auburn, NH, 2016).

______, 2016b, *The Periodic Table of the 192 Quarks and Leptons in The Theory of Everything: The U(4) Layer Group, Physics is Logic VI* (Blaha Research, Auburn, NH, 2016).

______, 2016c, *New Boson Quantum Field Theory, Dark Matter Dynamics, Dark Matter Fermion Layer Mixing, Genesis of Higgs Particles, New Layer Higgs Masses, Higgs Coupling Constants, Non-Abelian Higgs Gauge Fields, Physics is Logic VII* (Blaha Research, Auburn, NH, 2016).

______, 2016d, *Unification of the Strong Interactions and Gravitation: Quark Confinement Linked to Modified Short-Distance Gravity; Physics is Logic VIII* (Blaha Research, Auburn, NH, 2016).

______, 2016e, *MoND: Unification of the Strong Interactions and Gravitation II, Quark Confinement Linked to Large-Scale Gravity, Physics is Logic IX* (Blaha Research, Auburn, NH, 2016).

______, 2016f, *CQMechanics: A Unification of Quantum & Classical Mechanics, Quantum/Semi-Classical Entanglement, Quantum/Classical Path Integrals, Quantum/Classical Chaos* (Blaha Research, Auburn, NH, 2016).

______, 2016g, *GEMS: Unified Gravity, ElectroMagnetic and Strong Interactions: Manifest Quark Confinement, A Solution for the Proton Spin Puzzle, Modified Gravity on the Galactic Scale* (Pingree Hill Publishing, Auburn, NH, 2016).

______, 2016h, *Unification of the Seven Boson Interactions based on the Riemann-Christoffel Curvature Tensor* (Pingree Hill Publishing, Auburn, NH, 2016).

______, 2017a, *Unification of the Eleven Boson Interactions based on 'Rotations of Interactions'* (Pingree Hill Publishing, Auburn, NH, 2017).

______, 2017b, *The Origin of Fermions and Bosons,and Their Unification* (Pingree Hill Publishing, Auburn, NH, 2017).

______, 2017c, *Megaverse: The Universe of Universes* (Pingree Hill Publishing, Auburn, NH, 2017).

Eddington, A. S., 1952, *The Mathematical Theory of Relativity* (Cambridge University Press, Cambridge, U.K., 1952).

Fant, Karl M., 2005, *Logically Determined Design: Clockless System Design With NULL Convention Logic* (John Wiley and Sons, Hoboken, NJ, 2005).

Feinberg, G. and Shapiro, R., 1980, *Life Beyond Earth: The Intelligent Earthlings Guide to Life in the Universe* (William Morrow and Company, New York, 1980).

Gelfand, I. M., Fomin, S. V., Silverman, R. A. (tr), 2000, *Calculus of Variations* (Dover Publications, Mineola, NY, 2000).

Giaquinta, M., Modica, G., Souchek, J., 1998, *Cartesian Coordinates in the Calculus of Variations* Volumes I and II (Springer-Verlag, New York, 1998).

Giaquinta, M., Hildebrandt, S., 1996, *Calculus of Variations* Volumes I and II (Springer-Verlag, New York, 1996).

Gradshteyn, I. S. and Ryzhik, I. M., 1965, *Table of Integrals, Series, and Products* (Academic Press, New York, 1965).

Heitler, W., 1954, *The Quantum Theory of Radiation* (Claendon Press, Oxford, UK, 1954).

Huang, Kerson, 1992, *Quarks, Leptons & Gauge Fields 2^{nd} Edition* (World Scientific Publishing Company, Singapore, 1992).

Jost, J., Li-Jost, X., 1998, *Calculus of Variations* (Cambridge University Press, New York, 1998).

Kaku, Michio, 1993, *Quantum Field Theory*, (Oxford University Press, New York, 1993).

Kirk, G. S. and Raven, J. E., 1962, *The Presocratic Philosophers* (Cambridge University Press, New York, 1962).

Landau, L. D. and Lifshitz, E. M., 1987, *Fluid Mechanics 2^{nd} Edition*, (Pergamon Press, Elmsford, NY, 1987).

Misner, C. W., Thorne, K. S., and Wheeler, J. A., 1973, *Gravitation* (W. H. Freeman, New York, 1973).

Rescher, N., 1967, *The Philosophy of Leibniz* (Prentice-Hall, Englewood Cliffs, NJ, 1967).

Sagan, H., 1993, *Introduction to the Calculus of Variations* (Dover Publications, Mineola, NY, 1993).

Sakurai, J. J., 1964, *Invariance Principles and Elementary Particles* (Princeton University Press, Princeton, NJ, 1964).

Streater, R. F. and Wightman, A. S., 2000, *PCT, Spin, Statistics, and All That* (Princeton University Press, Princeton, NJ 2000).

Weinberg, S., 1972, *Gravitation and Cosmology* (John Wiley and Sons, New York, 1972).

Weinberg, S., 1995, *The Quantum Theory of Fields Volume I* (Cambridge University Press, New York, 1995).

Weinberg, S., 2000, *The Quantum Theory of Fields Volume III Supersymmetry* (Cambridge University Press, New York, 2000).

Weyl, H., 1950, *Space, Time, Matter* (Dover, New York, 1950).

Weyl, H., (Tr. S. Pollard et al), 1987, *The Continuum* (Dover Publications, New York, 1987).

INDEX

About the Author

Stephen Blaha is a well known Physicist and Man of Letters with interests in Science, Society and civilization, the Arts, and Technology. He had an Alfred P. Sloan Foundation scholarship in college. He received his Ph.D. in Physics from Rockefeller University. He has served on the faculties of several major universities. He was also a Member of the Technical Staff at Bell Laboratories, a manager at the Boston Globe Newspaper, a Director at Wang Laboratories, and President of Blaha Software Inc and of Janus Associates Inc. (NH).

Among other achievements he was a co-discoverer of the "r potential" for heavy quark binding developing the first (and still the only demonstrable) non-abelian gauge theory with an "r" potential; first suggested the existence of topological structures in superfluid He-3; first proposed Yang-Mills theories would appear in condensed matter phenomena with non-scalar order parameters; first developed a grammar-based formalism for quantum computers and applied it to elementary particle theories; first developed a new form of quantum field theory without divergences (thus solving a major 60 year old problem that enabled a unified theory of the Standard Model and Quantum Gravity without divergences to be developed); first developed a formulation of complex General Relativity based on analytic continuation from real space-time; first developed a generalized non-homogeneous Robertson-Walker metric that enabled a quantum theory of the Big Bang to be developed without singularities at $t = 0$; first generalized Cauchy's theorem and Gauss' theorem to complex, curved multi-dimensional spaces; received Honorable Mention in the Gravity Research Foundation Essay Competition in 1978; first developed a physically acceptable theory of faster-than-light particles; first derived a composition of extrema method in the Calculus of Variations; first quantitatively suggested that inflationary periods in the history of the universe were not needed; first proved Gödel's Theorem implies Nature must be quantum; provided a new alternative to the Higgs Mechanism, and Higgs particles, to generate masses; first showed how to resolve logical paradoxes including Gödel's Undecidability Theorem by developing Operator Logic and Quantum Operator Logic; first developed a quantitative harmonic oscillator-like model of the life cycle, and interactions, of civilizations; first showed how equations describing superorganisms also apply to civilizations. A recent book shows his theory applies successfully to the past 14 years of history and to *new* archaeological data on Andean and Mayan civilizations as well as Early Anatolian and Egyptian civilizations.

He first developed an axiomatic derivation of the forms of The Standard Model from geometry – space-time properties – The Extended Standard Model. It has a Dark Matter sector that approximates the ElectroWeak sector with Dark doublets and Dark gauge interactions. It also uses quantum coordinates to remove infinities that crop up in most interacting quantum field theories and additionally to remove the infinities that appear in the Big Bang and generate an inflationary growth of the universe. The Extended Standard Model has an ultra-high energy

GUT (Grand Unified Theory) limit with a U(4)⊗U(4) symmetry; and can be united with gravitation to form a Theory of Everything. (See *Physics is Logic Part II.*)

Blaha has had a major impact on a succession of elementary particle theories: his Ph.D. thesis (1970), and papers, showed that quantum field theory calculations to all orders in ladder approximations could not give scaling deep inelastic electron-nucleon scattering. He later showed the eigenvalue equation for the fine structure constant α in Johnson-Baker-Willey QED had a zero at $\alpha = 1$ not 1/137 by solving the Schwinger-Dyson equations to all orders in an approximation that agreed with exact results to 4^{th} order in α thus ending interest in this theory. In 1979 at Prof. Ken Johnson's (MIT) suggestion he calculated the proton-neutron mass difference in the MIT bag model and found the result had the wrong sign reducing interest in the bag model. These results all appear in Physical Review papers. In the 2000's he repeatedly pointed out the shortcomings of SuperString theory and showed that The Standard Model's form could be derived from space-time geometry by an extension of Lorentz transformations to faster than light transformations. This deeper space-time basis greatly increases the possibility that it is part of THE fundamental theory.Recently, Blaha showed that the Weak interactions differed significantly from the Strong, electromagnetic and gravitation interactions in important respects while these interactions had similar features, and suggested that ElectroWeak theory, which is essentially a glued union of the Weak interactions and Electromagnetism, possibly modulo unknown Higgs particle features, be replaced by a unified theory of the other interactions combined with a stand-alone Weak interaction theory. Blaha also showed that, if Charmonium calculations are taken seriously, the Strong interaction coupling constant is only a factor of five larger than the electromagnetic coupling constant, and thus Strong interaction perturbation theory would make sense and yield physically meaningful results.

In graduate school (1965-71) he wrote substantial papers in elementary particles and group theory: The Inelastic E- P Structure Functions in a Gluon Model. Phys. Lett. B40:501-502,1972; Deep-Inelastic E-P Structure Functions In A Ladder Model With Spin 1/2 Nucleons, Phys.Rev. D3:510-523,1971; Continuum Contributions To The Pion Radius, Phys. Rev. 178:2167-2169,1969; Character Analysis of U(N) and SU(N), J. Math. Phys. 10, 2156 (1969); and The Calculation of the Irreducible Characters of the Symmetric Group in Terms of the Compound Characters, (Published as Blaha's Lemma in D. E. Knuth's book: *The Art of Computer Programming Vols. 1 – 4*).

In the early 1980's Blaha was also a pioneer in the development of UNIX for financial, scientific and Internet applications: benchmarked UNIX versions showing that block size was critical for UNIX performance, developing financial modeling software, starting database benchmarking comparison studies, developing Internet-like UNIX networking (1982) and developing a hybrid shell programming technique (1982) that was a precursor to the PERL programming language. He was also the manager of the AT&T ten-year future products development database. His work helped lead to commercial UNIX on computers such as Sun Micros, IBM AIX minis, and Apple computers.

In the 1980's he pioneered the development of PC Desktop Publishing on laser printers. and was nominated for three "Awards for Technical Excellence" in 1987 by PC Magazine for PC software products that he designed and developed.

Recently he has developed a theory of Megaverses – actual universes of which our universe is one – with quantum particle-like properties based on the Wheeler-DeWitt equation of Quantum Gravity. He has developed a theory of a baryonic force, which had been conjectured many years ago, and estimated the strength of the force based on discrepancies in measurements of the gravitational constant G. This force, operative in 15-dimensinal space, can be used to escape from our universe in "uniships" which are the equivalent of the faster-than-light starships proposed in the author's earlier books. Thus travel to other universes, as well as to other stars is possible.

Blaha also considered the complexified Wheeler-DeWitt equation and showed that its limitation to real-valued coordinates and metrics generated a Cosmological Constant in the Einstein equations.

The author has also recently written a series of books on the serious problems of the United States and their solution as well as a book on the decline of Mankind that will follow from current social and genetic trends in Mankind.

In the past twelve years Dr. Blaha has written over 40 books on a wide range of topics. Some recent major works are: *From Asynchronous Logic to The Standard Model to Superflight to the Stars, All the Universe!, SuperCivilizations: Civilizations as Superorganisms, America's Future: an Islamic Surge, ISIS, al Qaeda, World Epidemics, Ukraine, Russia-China Pact, US Leadership Crisis,The Rises and Falls of Man – Destiny – 3000 AD: New Support for a Superorganism MACRO-THEORY of CIVILIZATIONS From CURRENT WORLD TRENDS and NEW Peruvian, Pre-Mayan, Mayan, Anatolian, and Early Egyptian Data, with a Projection to 3000 AD,* and *Mankind in Decline: Genetic Disasters, Human-Animal Hybrids, Overpopulation, Pollution, Global Warming, Food and Water Shortages, Desertification, Poverty, Rising Violence, Genocide, Epidemics, Wars, Leadership Failure.*

He has taught approximately 4,000 students in undergraduate, graduate, and postgraduate corporate education courses primarily in major universities, and large companies and government agencies.

The above paragraphs summarize much of his work over the past fifty years. This work is fully documented. He continues to engage in research and writing at Blaha Research.